Lecture Notes in Electrical Engineering

Volume 82

For further volumes:
http://www.springer.com/series/7818

M. Haykel Ben Jamaa

Regular Nanofabrics in Emerging Technologies

Design and Fabrication Methods for Nanoscale Digital Circuits

 Springer

M. Haykel Ben Jamaa
Commissariat à l'Energie Atomique
(CEA-DRT-LETI-DACLE-LISAN)
Rue des Martyrs 17
38054 Grenoble
France
e-mail: haykel.ben_jamaa@a3.epfl.ch

ISSN 1876-1100   e-ISSN 1876-1119

ISBN 978-94-007-0649-1   e-ISBN 978-94-007-0650-7

DOI 10.1007/978-94-007-0650-7

Springer Dordrecht Heidelberg London New York

Cover design: eStudio Calamar, Berlin/Figueres

Printed on acid-free paper

Springer is part of Springer Science+Business Media (www.springer.com)

Acknowledgments

I am thankful to Prof. Giovanni De Micheli and Prof. Yusuf Leblebici for their support of this work. I would like to dedicate this book to my family, especially to my parents.

Contents

List of Acronyms

1D	One-dimensional
2D	Two-dimensional
3D	Three-dimensional
AFM	Atomic force microscope
AHC	Arranged hot code
Al	Aluminum
ALD	Atomic layer deposition
As	Arsenic
ASIC	Application specific integrated circuits
Au	Gold
BGC	Balanced gray code
BHF	Buffered HF
BRC	Binary reflexive code
BTB	Band-to-band
CAD	Computer-aided design
Ce	Cerium
CF	Coupling fault
CG	Control gate
Cl	Chlorine
CLB	Complex logic block
CNT	Carbon nanotube
CNTFET	Carbon nanotube field effect transistor
Cr	Chromium
CMI	Center of micro- and nano-technologies [1]
CMOL	CMOS/molecular hybrid
CMOS	Complementary metal oxide semiconductor
CMP	Mechanical–chemical planarization or mechanical–chemical polishing
CVD	Chemical vapor deposition

DIBL	Drain induced barrier lowering
DRIE	Deep reactive Ion etching
E-beam	Electron beam
EDA	Electronic design automation
EDP	Energy-delay-product
EOT	Equivalent oxide thickness
EPFL	Ecole Polytechnique Fédérale de Lausanne, Swiss Federal Institute of Technology at Lausanne
EUV	Extreme ultraviolet
EUV-IL	Extreme ultraviolet interference lithography
F	Fluorine
FET	Field effect transistor
FIB	Focused ion beam
FO4	Fan-out-of-4
FPGA	Field programmable gate array
Ga	Gallium
GAOI	Generalized AOI, generalized AND-OR-Inverter
GC	Gray code
Ge	Germanium
GIDL	Gate-induced drain leakage
GNAND	Generalized NAND
GNOR	Generalized NOR
HC	Hot code
HMDS	Hexamethyldisilazane ($C_6H_{19}Si_2N$)
IC	Integrated circuit
In	Indium
IST	Iterative spacer technique
ITRS	International technology roadmap for semiconductors
Kr	Krypton
LB	Langmuir–Blodgett
LPCVD	Low pressure chemical vapor deposition
LTO	Low temperature oxide
LUT	Look-up table
MEMS	Microelectromechanical system
MOS	Metal oxide semiconductor
MOSFET	Metal oxide semiconductor field effect transistor
MSPT	Multi-spacer patterning technique
MVL	Multi-valued logic
MW	Mesowire
N	Nitrogen
Ni	Nickel
NIL	Nanoimprint lithography

NoC	Network-on-chip
NRC	n-Ary reflexive code
NW	Nanowire
O	Oxygen
OP	Operating point
OPC	Optical proximity correction
PANI	Polyanilin
PD	Pull-down
PDMS	Polydimethylsiloxane $[(C_2H_6OSi)_n]$
PEDAL	Planar edge defined alternate layer
PEO	Polyethylene oxide, polyethylene glycol $(C_{2n+2}H_{4n+6}O_{n+2})$
PG	Polarity gate
PLA	Programmable logic array
PMMA	Polymethylmethacrylate $[(C_5O_2H_8)_n]$
Poly-Si	Poly-crystalline silicon
Poly-SiNWFET	Poly-crystalline silicon field effect transistor
PR	Photoresist
PSF	Pattern sensitivity fault
Pt	Platinum
PU	Pull-up
Q-bit	Quantum bit
QCA	Quantum cellular automata
QWFET	Quantum well field effect transistor
RAM	Random access memory
RDR	Restrictive design rules
RET	Resolution enhancement techniques
RIE	Reactive ion etching
RTD	Resonant tunneling diode
S	Sulfur
Sb	Antimony
SB	Schottky barrier
SCE	Short channel effect
SEM	Scanning electron microscopy
SET	Single-electron transistors
Si	Silicon
SiNW	Silicon nanowire
SiNWFET	Silicon nanowire field effect transistor
SNAP	Superlattice nanowire pattern transfer
SoC	Systems-on-chip
SOI	Silicon-on-insulator
SRAM	Static random access memory
TC	Tree code
TEM	Transmission electron microscopy

Ti	Titanium
ULSI	Ultra large scale integration
VLS	Vapor–liquid–solid
VLSI	Very large scale integration
VPGA	Via patterned gate array
WPLA	Whirlpool PLA, Whirlpool programmable logic array

Reference

1. Center for Micro- and Nanotechnologies (CMI) at EPFL. http://cmi.epfl.ch

List of Symbols

Symbol	Definition	Unit
α	(i) Half-spacing between successive threshold voltages normalized to V_0	1
	(ii) Counter variable	1
β	Transistor gain factor $\mu C_{ox} W/L$	AV^{-2}
β_i	β at transistor i	
δ	Shift of any $V_{X,i}$ from $V_{A,i}$ normalized to V_0	1
ΔV_T	Distance between two successive $V_{T,i}$'s	V
$\delta(x)$	(i) Dirac distribution around x $= 0$	$\varnothing$
	(ii) Kronecker delta function	$\varnothing$
δI	Small signal of I	A
$\delta I_{d,i}$	Small signal of $I_{d,i}$	A
δI_u	Small signal of I_u	A
δV_{DS}	Small signal of V_{DS}	V
$\delta V_{DS,i}$	δV_{DS} at transistor i	V
δV_{DS}	Vector of $\delta V_{DS,i}$	V (vector)
δV_{GS}	Small signal of V_{GS}	V
$\delta V_{GS,i}$	δV_{GS} at transistor i	V
δV_T	Small signal of V_T	V
$\delta V_{T,i}$	δV_T at transistor i	V
δV_T	Vector of $\delta V_{T,i}$	V (vector)
ϵ	Test error probability	1
η	(i) Statistical factor $\eta = P_{nbr} \times P_{unq} \times P_{cnt} \times P_{int}$	1
	(ii) Sensitivity of V_{DS} to V_T in a decoder under test	1
η^{BU}	Statistical η for bottom-up technologies	1
η^{TD}	Statistical η for top-down technologies	1
Φ	Fabrication complexity $\Phi = \sum \phi_i$	1
ϕ_i	Number of photolithography/doping steps at MSPT step i	1
μ_α	Average number of code words having α identical digits	1
ν	Average number of patterns covered by a code word under defects	1

(continued)

(continued)

Symbol	Definition	Unit
Ω	Code space	$\varnothing$
Ω'	Addressable code space under defects	$\varnothing$
$|\Omega|_{\text{un}}$	Size of uniquely addressed code space	$\varnothing$
$|\Omega|_{\text{im}}$	Size of immune code space	$\varnothing$
Σ	Variability matrix	V^2 (matrix)
$\sigma(x)$	Sigmoid function of x	$\varnothing$
σ	Threshold voltage standard deviation ($= \sigma_{\text{T}}$)	V
$\sigma_{\delta\text{d},i}$	Standard deviation of $\delta I_{\text{d},i}$	A
$\sigma_{\delta\text{u}}$	Standard deviation of δI_{u}	A
$\sigma_{\text{d},i}$	Standard deviation of $I_{\text{d},i}$	A
$\sigma_{\text{d}}^{N_{\text{def}}}$	Standard deviation of $I_{\text{d}}^{N_{\text{def}}}$	A
σ_{T}	Standard deviation of V_{T}	V
σ_{u}	Standard deviation of I_{u}	A
τ	(i) Number of digits that flip in order to generate defect-induced noise (context of decoder design)	1
	(ii) Intrinsic delay of fan-out-of-1 inverter (context of ambipolar CNTFET)	s
$\bar{\tau}$	Mean value of number of digits τ	1
υ	(i) Normalized shift of V_{A} from middle of two successive threshold voltages normalized to V_0	1
	(ii) counter variable	1
A	(i) Pattern space	$\varnothing$
	(ii) Event of having $I_1 \leq I_{\text{s}}$	$\varnothing$
$\mathbf{A}$	First linearization matrix of the decoder	1 (matrix)
B	Event of having defect-induced noise, $B = \cup B_i$	$\varnothing$
B_i	Event of having exactly i nanowires generating defect-induced noise	$\varnothing$
$\mathbf{B}$	Second linearization matrix of the decoder	1 (matrix)
$\mathbf{a}$	Pattern	1 (vector)
$\mathbf{b}$	Pattern	1 (vector)
$\mathbf{b}^*$	Defective pattern	1 (vector)
C_{eff}	Effective crossbar (or memory) density	cm^{-2}
$C_{\text{eff}}^{\text{BU}}$	Effective crossbar (or memory) density in bottom-up technologies	cm^{-2}
$C_{\text{eff}}^{\text{TD}}$	Effective crossbar (or memory) density in top-down technologies	cm^{-2}
$\mathbf{c}$	Code word	1 (vector)
$\mathbf{c}^*$	Defective code word	1 (vector)
$\mathbf{c}^a$	Code word to the pattern a	1 (vector)
$\mathbf{c}^b$	Code word to the pattern b	1 (vector)
$\mathbf{d}$	Multi-digit error vector	1 (vector)
D	(i) Crosspoint density, memory density	cm^{-2}
	(ii) Elements of error subtree	$\varnothing$
D_{RAW}	Raw crosspoint (memory) density	cm^{-2}
D_{EFF}	Effective crosspoint (memory) density	cm^{-2}

(continued)

(continued)

Symbol	Definition	Unit
f	(i) Probability distribution function	$\varnothing$
	(ii) Photolithography half-pitch	nm
	(iii) Non-linear bijective application between N_D and V_T	$\varnothing$
f_u	Probability distribution function of useful signal	$\varnothing$
f_d^i	Probability distribution function of defect-induced noise generated by nanowire i	$\varnothing$
f_d	Probability distribution function of total defect-induced noise	$\varnothing$
g	Bijective application between pattern and V_T	$\varnothing$
g_{DS}	Output conductance of a FET	S
g_m	Transconductance of a FET	S
g_T	Sensitivity of I_{DS} to V_T: $\partial I_{DS}/\partial V_T$	S
h	Non-linear bijective application between elements of $\mathbf{D}$ and those of $\mathbf{P}$	$\varnothing$
$\mathbf{D}$	Doping matrix	cm^{-3} (matrix)
$\mathbf{I}$	Unit $M \times M$-matrix	1 (matrix)
I	Current through a nanowire under test	A
i	Counter variable	1
I_0	First thresholder parameter	A
I_1	Second thresholder parameter	A
$\bar{I}_{\delta d,i}$	Mean value of $\delta I_{d,i}$	A
$\bar{I}_{\delta u}$	Mean value of δI_u	A
I_d	Defect-induced noise in decoder under test	A
$I_{d,i}$	Defect-induced noise generated by nanowire i	A
$\bar{I}_{d,i}$	Mean value of $I_{d,i}$	A
$I_d^{N_{def}}$	Total defect-induced noise: $\sum I_{d,i}$	A
$\bar{I}_d^{N_{def}}$	Mean value of $I_d^{N_{def}}$	A
I_{ds}	Drain-source current	A
I_i	Intrinsic noise in decoder under test	A
$I_{i,0}$	Intrinsic noise generated by a single nanowire in decoder under test	A
I_{off}	Transistor off-current	A
I_{on}	Transistor on-current	A
I^{OP}	Operating point of I	A
I_s	Sensed signal in decoder under test $= I_u + I_d + I_i$	A
I_u	Useful signal	A
$\bar{I}_u$	Mean value of I_u	A
j	Counter variable	1
k	(i) Parameter of optimal-size multi-valued hot code	1
	(ii) Counter variable	1
$\mathbf{k}$	Parameter vector of multi-valued hot code	1 (vector)
L	Transistor length	nm
L_l	Photolithography pitch	nm
L_n	Sub-photolithographic pitch, nanoscale pitch	nm

(continued)

(continued)

Symbol	Definition	Unit
M	Length of code word, number of doping sequences in a nanowire (equivalent definitions)	1
N	Number of nanowires in a contact groups	1
N_D	Doping level (donors or acceptors)	cm^{-3}
N_{def}	Number of nanowires generating defect-induced noise in a decoder under test	A
N_{off}	Number of non-activated nanowires in a decoder under test	A
N_{use}	Number of useful nanowires in a decoder under test ($N_{use} = 0$ or 1)	A
n	Logic value	1
$\mathbf{P}$	Pattern matrix	1 (matrix)
P_0	Probability that no nanowire is addressed	1
P_1	Probability that 1 nanowire is addressed	1
P_2	Probability that ≥ 2 nanowires are addressed	1
P_{cnt}	Probability of a good nanowire control	1
$P_{contact}$	Probability of a good nanowire ohmic contact	1
P_{int}	Probability of no nanowire loss at the interface between contact groups	1
P_{nbr}	Probability of a non-broken nanowire	1
P_{unq}	Probability of a unique nanowire	1
p_I	Probability of a type-I error in a code space	1
p_{II}	Probability of a type-II error in a code space	1
p_α	Probability that μ_α code words undergo error sequences that make them covered	1
p_d	Probability of a flip-down error in a code word	1
p_{im}	Probability of an immune code space	1
p_u	Probability of a flip-up error in a code word	1
p_U	Probability of a uniquely covered code space	1
q	On/off current ratio to detect a digit	1
r_i	$R_M \| g_{m,i}$	Ω
R_M	Resistance of the nanowire memory part	Ω
S	(i) Set of all (p_α, μ_α)	$\varnothing$
	(ii) Set of indexes of digits that undergo flip-ups	$\varnothing$
$\mathbf{S}$	Step doping matrix	cm^{-3} (matrix)
s	Sequence digit shifts to generate defect-induced noise	1 (vector)
T	Error transformation matrix	$\varnothing$
$\mathbf{t}^i$	Error transformation affecting digit level i	$\varnothing$
$\mathbf{V}$	Threshold voltage matrix	V (matrix)
$\mathbf{v}$	Vector with all entries set to 1	1 (vector)
V_+	V_{PG} for n-type device	V
V_-	V_{PG} for p-type device	V
V_0	(i) Decoder normalization voltage, set to V_{DD} (context of decoder test)	V

(continued)

(continued)

Symbol	Definition	Unit
	(ii) V_{PG} for a low-conductivity transistor (context of ambipolar CNTFET)	V
V_A	Applied voltage at the decoder (gate-to-ground voltage)	V
$V_{A,i}$	V_A at transistor i	V
V_{CG}	Control gate voltage (ambipolar CNTFET)	V
V_{DD}	Supply voltage	V
V_{ds}	Drain-source voltage (context of measurement)	V
V_{DS}	Drain-source voltage (context of testing)	V
$V_{DS,i}$	V_{DS} at transistor i	V
V_{gs}	Gate-source voltage (contest of measurement)	V
V_{GS}	Gate-source voltage (context of testing)	V
$V_{GS,i}$	V_{GS} at transistor i	V
V_P	Power supply of decoder under test	V
V_{PG}	Polarity gate voltage (ambipolar CNTFET)	V
V_{SS}	Ground, reference voltage	V
V_T	Threshold voltage	V
$\overline{V}_T$	Mean value of V_T	V
$\overline{\mathbf{V}}_T$	vector of $\overline{V}_{T,i}$	V (vector)
$\overline{V}_{T,i}$	$\overline{V}_T$ at transistor i	V
$\mathbf{V}_T$	Vector of $V_{T,i}$	V (vector)
$\mathbf{V}_T^{OP}$	$\mathbf{V}_T$ at operating point	V (vector)
$V_{T,i}$	V_T at transistor i	V
$V_{X,i}$	Threshold voltage to detect digit i	V
V_{Tn}	Threshold voltage of n-type device	V
V_{Tp}	Threshold voltage of p-type device	V
W	Transistor width	nm
Y	Statistical yield	1

Chapter 1
Introduction

Modern societies have been deeply reshaped by a sequence of industrial revolutions. Some of them were completed and they represent today a chapter of history, while some others are still evolving and generating more changes in the societies. One fundamental aspect that characterizes modern societies is the increasing level of usage of electronic systems that are partly becoming indispensable for some daily activities. Conversely, the daily activities are reshaped according to the new systems uninterruptedly coming to the market. This tight link between the needs of modern societies and the electronics system represents one of the latest revolutions.

Electronic systems are complex systems generally conceived around a physical core, the *hardware*, which can be customized and programmed by adding soft modules called *software*. The hardware of electronic systems is versatile. This diversity concerns the way it is fabricated, the *manufacturing* or *fabrication technology*, and the way the available manufacturing technology is utilized to design the full system, i.e., the *design technology*. Moreover, the electronic systems are segmented in terms of usage or *application*: while some systems are used for communication, transportation or office applications, some others are utilized for entertainment purposes, or in more vital fields such as health care, energy, security and defense.

It is interesting to notice that the hardware of all these diverse and numerous systems is ultimately a complex *integrated circuit* (*IC*), whose manufacturing and design make the difference. An integrated circuit is a set of basic elements that control the flow of electrical current in a defined way in order to perform a certain set of operations. These basic elements are called *transistors*. The invention of such devices goes back to the first half of the twentieth century. But their assembly into integrated systems is just less than half a century old.

After the transistor was invented, the scientific community rapidly became aware of the potential benefits of this novel technology. As a matter of fact, transistors are made of commonly available materials: namely germanium (Ge),

M. H. Ben Jamaa, *Regular Nanofabrics in Emerging Technologies*,
Lecture Notes in Electrical Engineering, 82, DOI: 10.1007/978-94-007-0650-7_1,
© Springer Science+Business Media B.V. 2011

and especially silicon (Si). In addition, they are able to maintain the ability to control the current switching on and off despite the scaling down of their size. Then, by scaling the device and using a larger number of transistors, the integrated circuits had the potential of becoming more complex and performing more functions. The industrial community did not overlook this unique opportunity, and it initiated the era of *semiconductor industry*.

The growth of the semiconductor industry in the last half-century has not been as sensitive to the cyclicity of the world economy as other markets have, leading to a continuous growth of the developed systems from small integrated circuits made of a few transistors to a *very large scale integration* (*VLSI*) and *ultra large scale integration* (*ULSI*) levels with millions and billions of transistors. Many novel techniques are being investigated today in order to keep this growth. These techniques are the results of a wide range of research efforts in different fields, including both the manufacturing technologies (processing and device engineering) and the design technologies (at the circuit and system levels). The previous decades of research have been mostly focusing on the scaling down of the device features in order to further increase the integration level, while roughly keeping the same underlying transistor physics and the related system paradigms unchanged.

The *international technology roadmap for semiconductors* (*ITRS*) [1] is a consortium of leaders in the fields of semiconductor research and industry, whose goal is to survey the trends of the semiconductor technology and predict its future evolution. Today, the ITRS recognizes the existence of physical limits to this growth: the electronics-based technologies cannot be scaled down beyond certain dimensions that are defined by some physical limits [2].

The semiconductor technology that dominates the electronics market today is the *complementary metal-oxide-semiconductor (CMOS)* technology, which is based on the utilization of complementary transistors, designated by n- and p-types, which carry, respectively, electrons and holes. The challenging task of the ITRS today is to find a way to continue the scaling of CMOS technology or its fundamental replacement by other technologies promising more scaling opportunities. The efforts undertaken in this sense are reflected by the variety of processing techniques, device architectures and system designs that have been investigated in the last decade.

The candidates that are viable for a possible replacement of CMOS technology are commonly called *emerging technologies*. These tentative solutions are based on novel materials, device physics, circuit designs etc. They share some common aspects: for instance, the dimension scaling is pushed so far, that typical device dimensions are in the range of an average-size molecule. At this scale, the uncertainty becomes high and the variability increases at the single device level, so that a reliable operation of the system cannot be guaranteed anymore. On the other hand, the accurate placement of devices with the size of single molecules is challenging the manufacturing and increasing the overall variability.

The increasing variability and limited manufacturing abilities set a limit on the success of the industrial exploitation of electronic systems. It is therefore necessary to find technological solutions that enable a large scale implementation of emerging technologies. One possible way to survive with an unreliable technology is to increase the redundancy, i.e., to use more instances of the same device or unit, in such a way that any failing part of the circuit can be replaced. Another way is to arrange devices and units in a regular way in order to increase the determinism of the circuit. Regularity is often compatible and desirable in redundant systems.

Regular architectures in emerging technologies are the driver of this research work. The global objective of this work is, first, to assess a set of emerging technologies, then to address some challenges arising from the manufacturing level at the design level. Consequently, the work is presented across different design fields and manufacturing technologies.

This work is organized across different technologies defined by two types of devices and it focuses on their integration into large systems. The considered technologies are defined by *one-dimensional* (*1-D*) devices: *silicon nanowires* (*SiNWs*) and *carbon nanotubes* (*CNTs*), both of which are promising a better performance compared to CMOS and more opportunities to further scale down the dimensions. The system integration starts from the fabrication and goes to the system level: for every one of the two types of devices, this research work will demonstrate the possibility of addressing a complex issue related to the fabrication by applying relatively simple solutions on the design level.

Both SiNW and CNT technologies are preferably locally arranged into regular and parallel layers, even though the global circuit can be defined in a custom design way. But their intrinsically different operation, physics and fabrication techniques are the reasons for the different challenges facing their integration into large scale systems.

When it comes to SiNW technology, the regular arrangement of devices into arrays, called *crossbars*, embodied within a surrounding CMOS part, is the most promising architecture. However the link between the crossbars and the CMOS part is a challenging task. In this work, a demonstration of a crossbar fabrication technique that can be potentially integrated in an easy way into a CMOS process is performed. Then, solutions are given in order to link the crossbar to the CMOS part. These solutions include the choice of the nanowire codes, the design of the decoder that links both parts and performing a reliable decoder testing operation.

On the other hand, among the most important challenge that CNT technology is facing is the doping of *CNT field effect transistors* (*CNTFETs*). As explained before, both n- and p-type devices may be needed in the most common circuit device approaches. However, for CNT technology, n-type devices are difficult to obtain through usual technique, called chemical doping in CMOS technology. Recently, it has been demonstrated that the polarity of CNT devices, i.e., whether they are n- or p-type devices, can be controlled during the operation of the device. The device is called in this case *ambipolar*, i.e., operating as n- and p-device at the same time. This represents an opportunity for new design methodologies for *ambipolar circuits*, which will be introduced in this work and their benefit will be assessed.

The novelty and contributions of this work consists in its multidisciplinary approach, in which the system design is enhanced by leveraging some opportunities offered by the versatile fabrication technologies. The complementary point of view is also a novel contribution of this work: i.e., some challenges of the fabrication technology are addressed at the system level.

This chapter is an introduction organized the following way. First, the past decades of linear scaling and the difficulties facing CMOS technology at the latest milestones are surveyed. Then, promising emerging technologies that may either sustain CMOS technology or create technological discontinuities are presented. Thereafter, some promising regular architectures are surveyed and the challenges that they are facing are explained. This chapter is concluded with the contributions and limitations of this work, and by explaining its outline.

1.1 The Linear Scaling

The evolution of the semiconductor industry is among the most fascinating industry stories in the twentieth century. It is one of the latest industrial revolutions, whose impact can be seen in most of the details of our life today. Our basic needs have been reshaped in the last fifty years or more in such a way that electronic elements performing computation, storage, sensing, connectivity, entertainment, security, health care, and much more, are hardly separable from our daily activities. The range of applications in which microelectronics products are present is becoming even larger everyday. It is remarkable that most of the semiconductor systems are built out of the same fundamental element, which is the transistor.

The history of the transistor, driving the whole semiconductor industry and reshaping the needs of modern societies, does not go more than 60 years back in time. The exact date of birth of the transistor is not uncontroversial, since a non-advertised patent for the *field effect transistor* (*FET*) was filed in the twenties [3], but never used for any practical implementation by its inventor, until it became the basis of the first transistor fabricated in 1947 by researchers from Bell Labs [4] and then the first device that became known as the bipolar junction transistor was invented [5]. The first fabrication of a FET was successfully performed in 1959 [6], which evolved into the *metal-oxide-semiconductor field effect transistor* (*MOSFET*) we know today.

The evolution of integrated circuits is captured by Moore's law [7], which was stated in 1965, predicting an exponential growth of the number of transistors per die every year. Even though the growth rate was not constant during the last decades, still it was about a factor of 2 every 2 years, in the same range as predicted by Moore's law initially Fig. 1.1. This roughly constant growth is unique among all industries and raised the question about the origins and the reasons of this sustained growth. A naïve answer that has been continuously reported states the cost reduction. This used to be true in early ages of the semiconductor industry. Today, the silicon costs are marginal, and the *electronic design automation* (*EDA*) tools, mask fabrication and circuit design costs are

Fig. 1.1 More's law: transistor count in released processors doubles approximately every 2 years. This is equivalent to saying that intrinsic speed $1/\tau$ improves by 19%Year. Historically, the improvement was slower: $\sim 17\%$Year

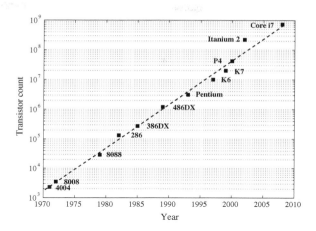

becoming tremendous compared to material costs. The emotional answer is tightly linked to the enthusism driving the research in the semiconductor field: it is always challenging and curious to assess the limits and see how far research can go with the available knowledge and come up with novel solutions. The most pragmatic answer comes from the fact that Moore's law represents a roadmap for future achievements, which is practically a non-binding agreement about the future between the actors of the semiconductor industry. Having a universal agreement between competitors is a highly desirable situation for semiconductor leaders because it decreases the uncertainty about the future of the semiconductor market. Such an agreement may be considered as a collusion though, which is a forbidden marketing practice in an economy with a perfect competition. Nevertheless, the semiconductor market has the unique chance of having its roadmap established and purposefully sustained by the competitors without any unlawful agreement. The second marketing-related answer is the continuous need for new products in a consumer-driven world economy. Since packaging makes the ICs morphologically identical, new semiconductor products have generally been differentiated by the device integration level. Recently, marketing experts have discovered that besides the hardware level, the application level is a strategic way to differentiate products as well [8].

The last decades, in which sustaining Moore's law was straightforward, are commonly called the happy-scaling period. During this period, the device dimensions and voltages have been scaled by a factor of α periodically, which results in a global improvement of the device density and reduction of the device power consumption by a factor of α^2, and a speed-up of the intrinsic device by a factor of α, as shown in Table 1.1 [9]. In order to sustain this trend, the ITRS [1] was founded as a consortium representing the leading actors in the field of semiconductors, which periodically publishes predictions for the evolution of the semiconductor industry, including device research, process integration, semiconductor materials, circuit design, interconnect issues, packaging, EDA tools...

Table 1.1 Linear scaling rules

Parameter	Scaling factor
Transistor length and width (L, W)	$1/\alpha$
Junction depth (x_j)	$1/\alpha$
Oxide thickness (t_{ox})	$1/\alpha$
Doping concentration (N_d, N_a)	α
Supply voltage (V_D)	$1/\alpha$
Drive current (I_D)	$1/\alpha$
Electric field (E)	1
Capacitance ($\epsilon \cdot A/t_{ox}$)	$1/\alpha$
Delay time ($\tau = C \cdot V_D/I_D$)	$1/\alpha$
Power dissipation (roughly $V_D \cdot I_D$)	$1/\alpha^2$
Device density ($\sim 1/A$)	α^2

1.2 The Latest Milestones

The linear scaling is expected to reach the 22-nm technology node. However, it is facing many challenges today, and the vision beyond this point is not a matter of a common consensus. This section surveys the most important issues that need to be addressed with the latest milestones at different research levels.

1.2.1 Fabrication Technology

Until approximately the 90-nm technology node, the overall IC design flow was relying on the linear scaling of devices that was requiring a corresponding scaling of the design rules in order to insure the feasibility of the fabrication. Today, the design rules do not scale in a similar way, and even the simplest layout geometries may tend to create patterns that cannot be reliably printed with the available photolithographic techniques. The manufacturing is becoming more challenging, and it makes the design more difficult as well.

In the past, photolithographers applied an aggressive scaling of the wavelength of light in order to achieve the required scaled features. The difficulty in developing fabrication equipments that are compatible with shrinking wavelengths of light, obliged the manufacturing community to use larger wavelengths than the feature size Fig. 1.2. Thus, it became necessary to introduce aggressive techniques that sustained the scaling of the dimensions, while using the same light source [10]. Most of these alternative approaches use post-tapeout manufacturability enhancement techniques. One of the available options is a set of strong *resolution enhancement techniques* (*RETs*), such as phase-shifting masks [11]. These techniques use the properties of the optical projection process to reduce the resolution features and the sensitivity of the printed shapes to process variation during the photolithography steps. Another set of tools is the *optical proximity correction*

Fig. 1.2 Lithography
roadmap: while the feature
size is scaled down, the
lithography wavelength does
not scale accordingly. This
situation appears at the
130-nm node. The
introduction of EUV light in
lithography becomes
necessary

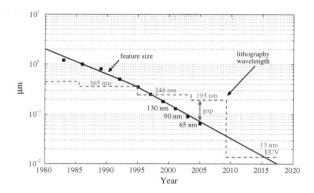

(*OPC*) [11], which attempts to compensate for systematic shape distortions of
fabricated devices by applying the inverse distortions when converting the design
shape (tapeout) to mask shapes.

The OPC technique is a data processing step with a tremendous cost in terms of
time resulting from the complicated computation required in order to calculate the
inverse distortions. On the other hand, although RETs can be used in order to
improve the image quality of some patterns in the layout, they tend to compromise
the quality of other patterns. There is a set of non-RET compliant patterns that
need to be identified and avoided by the designer during the mask drawing steps.
However, it is difficult to assess all non-RET compliant patterns because of the
large possible combination of layout patterns that exist in an average library of
hundreds of standard-cells.

The link between the design and manufacturing communities is becoming more
critical than it used to be in the past [12]. This link can be summarized in a set of
design rules that designers have to respect when they create an IC; such design
rules insure the manufacturability of the layout. Their number is highly increasing
with each technology node, making the task of the designer more difficult and
time-consuming. On the other hand, the definition of the design rules becomes a
difficult, sometime impossible task for the manufacturing team, since these rules,
that used to reflect the mechanical, optical and chemical processes controlling the
photolithography, are becoming design-dependent. It is therefore impossible to
validate a process through a universal set of design rules, because of the geometry
pattern explosion. Since design rules can be only validated for a given design, or
more precisely for a given set of patterns, designs with non validated patterns are
fault-prone and they dramatically impact yield; thus, representing a major concern
of the semiconductor industry.

The need for limiting the pattern explosion sets new methodologies when it
comes to defining the design rules. The general tendency is towards more regularity of the permitted patterns [12], corresponding to the set of patterns that were
tested by the manufacturing team and validated as RET-compliant. A set of
restrictive design rules (*RDR*), which can be also seen as radical design rules, has
been recently proposed by different semiconductor manufacturers [11, 13] with

two goals: first, to ensure an RET-compliance of the design, and second, to reduce the computational complexity of the OPC processing step. The main concepts introduced by RDRs are: the use of smaller-and possibly discretized-range of line widths for critical patterns, the unidirectionality of critical features, such as gates, the placement of critical features on a discretized grid, and the limitation of the combinations of proximities for critical features.

1.2.2 Device Design

1.2.2.1 Electrostatic Channel Control

With the channel scaling, short channel effects are becoming more severe Fig. 1.3. The *short channel effect (SCE)*, measured at $V_{ds} = 0$ V, reflects the lowering of the threshold voltage V_T with a decreasing channel length, due to a *two-dimensional (2-D)* distribution of the surface potential in the channel. Moreover, in short channel devices, the *drain-induced barrier lowering (DIBL)* consists in making the threshold voltage dependant on the drain bias V_{ds} due to the penetration of the drain depletion zone deep into the short channel. Both CSE and DIBL induce a lowering of V_T (for n-type transistors), which makes the devices more vulnerable to variability. DIBL has in addition critical consequences on the circuit design, due to the fact that the saturation of the I_{ds}-V_{ds} curve does not occur any more, and the device on- and off-states (i.e., $V_{gs} > V_T$ or $V_{gs} \leq V_T$) can be unintentionally set by V_{ds}.

1.2.2.2 Carrier Mobility

Setting the threshold voltage properly and providing a better control of the short channel effects necessitate the channel doping level to be increased to extremely

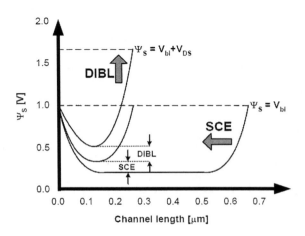

Fig. 1.3 SCE and DIBL for different channel lengths and supply voltages [8, 20]

higher values than the theoretical scaled values given by the linear scaling theory (Table 1.1). As a result of the high channel doping level, the carrier mobility in the channel deteriorates [14], causing a lower drive current and a slower device. The bulk mobility values for holes ($\sim$500 cm^2/V s) and electrons ($\sim$1,400 cm^2/V s) are therefore rarely measured in the silicon channel because of the presence of dopants; and the situation worsens with the applied electrical field, due to phonon scattering [15], and by the scattering at the Si/SiO$_2$ interface between the channel and the gate oxide [16].

1.2.2.3 Leakage

The need for an extremely high valued of channel doping causes the increase of the junction leakage due to *band-to-band* (*BTB*) tunneling and *gate-induced drain leakage* (*GIDL*), both kinds of leakage leading to higher power consumption [17, 18].

On the other hand, the scaling of the oxide thickness along with the other device dimensions provokes an exponential increase of the gate leakage, occurring as a tunnel current through the gate oxide, with the linear decrease of the gate thickness [19].

1.2.2.4 Gate Stack

Historically, poly-Si was used for the gate stack because of its easy integration into CMOS processes. However, several issues are arising today due to the use of poly-Si gates. As a matter of fact, by applying the gate voltage, the current carriers are depleted at the gate to gate oxide interface. On the other side of the gate oxide barrier, namely at the channel to gate oxide interface, a dark space is created and is due to charge quantization effects in the inversion layer. Both dark space and poly-depletion thickness act as additional insulators between the channel and the gate, by virtually increasing the *equivalent oxide thickness* (*EOT*), as depicted in Fig. 1.4.

This problem is today addressed by using a metal gate, since in metals no carrier depletion occurs. On the other hand, metal gates have the other advantage of being suitable for high-κ materials used as gate insulators in order to reduce the gate leakage.

1.2.2.5 Access Resistance

In addition to the channel resistance, a certain number of access resistances control the device performance. These include the contact resistance, drain/source resistance and the overlap resistance defined by the path underneath the gate and its spacers. It is very difficult to keep the access resistances constant with the decreasing

Fig. 1.4 Equivalent oxide
thickness: impact of gate
depletion and dark space
[8, 20]

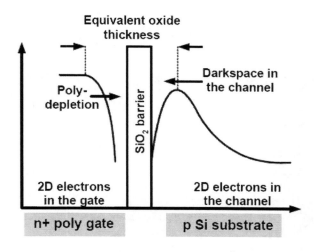

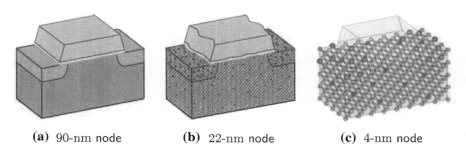

(a) 90-nm node   (b) 22-nm node   (c) 4-nm node

Fig. 1.5 Evolution of device modeling from continuous to discretized concepts [22]: **a** Dopant fluctuation canceled by continuum at 90-nm node. **b** At 22-nm node: only 50 Si atoms along the channel. **c** At 4-nm node: discretized distribution of dopants is source of high variability

device dimensions, mainly because of limited doping control in the drain and source implantation regions [21].

1.2.2.6 Variability

Shrinking the channel dimensions results in a lower total number of dopants in the channel, which in turns amplifies the variability of the number and location of doping species inside the channel. Under these condition, statistical averaging over the whole set of doping species becomes impractical, and the threshold voltage consequently suffers from a higher variability. The Fig. 1.5 from [22] illustrates the impact of the shrinking dimensions on the fluctuation of dopant distribution along the channel.

1.2.3 System Design

Many device parameters do not scale according to the linear scaling theory [9] as previously explained. As a result, many device parameters that used to be fixed by the technology node are becoming design parameters, such as the supply and the threshold voltage. The designer can optimize these parameters in order to obtain the best trade-off in terms of area, power and delay.

While the improvement in circuit density and performance has been achieved in the last decades by scaling down the transistors, latency of on-chip wires does not follow the same trend. The latencies of local interconnects scale accordingly; but global signals represent the big challenge. Their relative length to the device size does not scale down nor does is remain constant, because the chip size does not decrease while moving towards future technology nodes, given the fact that the number of transistors per chip continuously increases, and the die area roughly remains constant. The delay of global signals increases quadratically with their length, and their bandwidth accordingly degrades [12]. Wider interconnect wires may improve the overall latency, however they have a high cost in terms of energy-per-bit and bandwidth density (bits per second per unit routing width). Therefore, buses are becoming a less attractive option for on-chip connectivity; and they have to progressively be replaced by on-chip networks and other on-chip communication systems [23].

The clock distribution remains an important challenge in the design of digital systems. Most of modern clock systems are based on a single clock phase that is distributed globally, while local clock phases may be generated locally using different clock conditioning circuits. With the increasing relative length of global signals to the critical dimensions, the task of distributing the clock signal simultaneously (i.e., no skew) and periodically (i.e., no jitter) everywhere is becoming increasingly challenging. The relative non-scaling of wire delay and the increasing amount of capacitance per unit area worsen the clock latency (Fig. 1.6 from [12])

Fig. 1.6 Global clock latencies from published microprocessor designs [12]

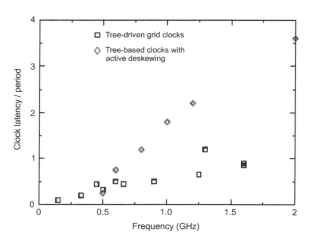

and increase the clock gain (measured as the ratio of the clock load capacitance to the capacitance driven from the phase-lock-loop reference) to extremely high values (more than 10^5). Moreover, skew and jitter become difficult to control when the temperature, process and supply voltage variation temporally and spatially increase.

1.2.4 Design Tools

Delay and power used to be the most important system design parameters. However, the increasing variability at the device level has noticeable consequences on the yield of the whole system. Starting from approximately the 65-nm technology node, *computer-aided design* (*CAD*) tools had to be enhanced or rethought in order to enable a yield-aware design approach. There are two types of yield losses: catastrophic and parametric losses. The first category represents the loss due to physical and structural defects, e.g., opens, shorts, etc; while the second category includes all sorts of losses due to variation in process parameters, such as the threshold voltage, the gate oxide thickness, etc. The Fig. 1.7 from [12] illustrates the increase in I_{off} in actual devices compared to ideal ones due to the use of subwavelength lithography.

The simplest performance modeling is based on linear approximations. However, such models are not able to accurately capture the variation in large systems. Quadratic approximations are therefore needed, which increases the complexity of the design tools as well as the simulation time drastically. In a 65-nm CMOS process, there are typically 100 variable process parameters, which result in about 10^4 parameters to be included in the modeling tool. The simulation time increases drastically with this large number of parameters to be captured by the CAD tool. In order to capture statistically rare events, the number of simulation has to be increased. Such events are very important to capture: even though they may rarely occur in a single device or basic bloc, they generally occur with a high probability in

Fig. 1.7 Due to subwavelength lithography, actual devices are more leaky than ideal ones [12]

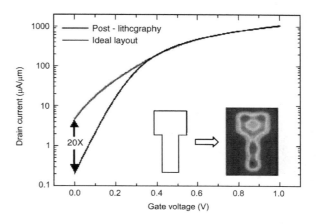

large arrays of such devices or basic blocs, and they may impact a large part of the circuit. For instance, the *static random access memory (SRAM)* is highly sensitive to statistically rare events.

1.3 Emerging Technologies

Emerging technologies designate in general non-CMOS devices and systems. In this context, CMOS is to be understood in a broader way than just the bulk CMOS technology: i.e., it is considered to be the set of all technologies with a silicon channel independently of the substrate type, Si or *silicon-on-insulator (SOI)*, and the device architecture (single or multi-gate, planar or 3-D gate...), provided that the underlying physics is based on the traditional drift-diffusion current of holes or electrons set by the gate-controlled enhancement, depletion or inversion of the channel. Emerging technologies are divided into two sets of technologies with respect to CMOS. First, we distinguish those that are based on a similar physics-within a certain limit and at a certain abstraction level-and that are expected to extend CMOS technology, such as CNT and SiNW technologies. Then, there is a set of emerging technologies that are based on different physics in order to avoid to face the same physical limitations that are challenging CMOS technology. The different physics is essentially characterized by a different state variable, for instance the spin, the molecular state or the *quantum bit (q-bit)*, instead of the charge used in CMOS technology. Such emerging technologies include for instance *quantum cellular automata (QCA)*, molecular electronics, spinotronics, quantum computing, and many others (Fig. 1.8 from [1]).

Fig. 1.8 Taxonomy for emerging technologies [1]

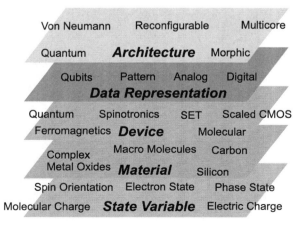

1.3.1 Selecting Emerging Technologies

Emerging technologies are evaluated in terms of their ability to replace CMOS. Many criteria are considered during the evaluation of emerging technologies: for instance, their technological integration into the already exiting infrastructural platform, their performance, their ability to scale down... The ITRS committee continuously surveys the emerging technologies and carries out their evaluation with respect to CMOS technology. Today, the general tendency is to believe that until 2015, none of the emerging technologies being currently investigated will have the potential to replace scaled CMOS. Therefore, the large amount of research activity has been focusing on the question of whether there are any novel information processing paradigms in which the emerging technologies can perform better than CMOS. Such information processing paradigms shall be different than the traditional Boolean computation that has been dominating in the last half-century.

Consequently, the research in emerging device technologies is highly related to the overall device organization and system architecture. The industrial trend today is to support multi-core architectures with heterogeneous computational elements. Efficient on-chip networks connect the different cores with another independently of the core type. This trend is consistent with the research activity towards dedicated coprocessor utilizing novel devices for specialized applications, embedded into a CMOS-based system. Such applications include for instance image recognition, speech recognition and data mining; and may be biology-inspired instead of operating like Boolean systems.

1.3.2 CMOS Extensions

The extension to CMOS is the goal of a large set of emerging technologies dedicated to the challenges facing the device channel. In order to improve the mobility and electrostatic control, novel materials are being introduced. There are two possible approaches to replace the channel material. In the first approach, novel low-dimensional devices such as SiNWs, CNTs and graphene nanoribbons are replacing the Si channel. In the second approach, high mobility compounds are used in order to enhance the channel mobility. These compounds include Ge and III-V compound layers.

1.3.2.1 Low Dimensional Structures

The generic expression "low dimensional structures" is an approximative description of CNTs, SiNWs and graphene nanoribbons. Actually, CNTs and graphene nanoribbons are two-dimensional carbon sheets, which are rolled for

CNTs and flat for graphene nanoribbons; while SiNWs are in reality three-dimensional structures with extremely small cross-section dimensions. In all these cases, the confinement effects are so noticeable, that the structures behave in an inherently different way compared to the bulk material, and they can be therefore considered as quasi one-dimensional structures. They are expected to have a much higher mobility than bulk silicon in MOSFETs due to their confinement, which makes them attractive candidates to form the transistor channel in emerging devices. However, there are still many challenges facing the fabrication and assembly of these structure, the fabrication of single transistors, and the large scale integration of the technology.

When it comes to CNTs (Fig. 1.9a), one of the most challenging aspects is the control of the structure chirality, which determines whether it has a semiconducting or a metallic behavior, i.e., whether it can be used as a channel in a FET or not. It has been recently reported in literature that accurate control of the plasma parameters and optimized engineering of the catalyst during the CNT growth phase helps gain an improved control over the CNT chirality [24, 25]. Alternatively, selective etching of metallic CNTs after the growth phase is also possible, and it helps better control the chirality of the fabricated structures [25]. Another on-going research topic is the large scale doping of CNTs. Even though single structures could be n- and p-doped, wafer scale doping has been demonstrated only with a p-type behavior [26]; and some research groups have suggested the use of intrinsic CNTs [27–29] in *Schottky barrier* (*SB*) devices. The assembly of CNTs into parallel arrays has successfully been demonstrated on a wafer scale [26, 30].

The fabrication of SiNWs (Fig. 1.9b) has been reported in literature with various techniques, and will be surveyed in Sect. 2.1. The doping and electrical behavior are well controlled by the fabrication process, compared to the case of CNTs [31–33]. The arrangement of SiNWs has been demonstrated in parallel layers and in three-dimensional stacks [34–40]. However the placement of single nanowires in specific locations has been demonstrated only for a limited set of fabrication techniques. The issue of placing nanowires will be surveyed in detail in Sect. 2.2. SiNWs offer the unique opportunity, compared to bulk silicon devices, to wrap the gate around the nanowire, thus guaranteeing a better electrostatic control of the current through the nanowire [41].

(a) Carbon nanotube   **(b)** Silicon nanowire   **(c)** Graphene nanoribbon

Fig. 1.9 Low dimensional structures as channel replacement: **a** CNTs are rolled graphene sheets. **b** SiNWs are Si atoms in a bulk arrangement, having variable length and cross-section dimensions (shown here with contacts). **c** Graphene nanoribbons are narrow graphene stripes

Graphene sheets (Fig. 1.9c) have been gaining more interest because of the high current carrier mobility through them, which can approach 10^4 cm^{2}V^{-1}s^{-1}. The confinement of this material in graphene nanoribbons opens the bandgap and makes the structures semiconducting and highly interesting as a channel for FETs.

1.3.2.2 High Mobility Materials

This approach is similar to the previous one in the sense that silicon in the channel is replaced by a material that promises a higher mobility. However, the material of the device channel is confined only in the vertical direction to the substrate, unlike the previous approaches in which CNTs, SiNWs and graphene nanoribbons are confined in both directions perpendicular to the channel axis. Therefore, the channel considered in this case represents a quantum well, mostly fabricated with Ge or III-V compounds with a narrow bandgap and a higher electron mobility than silicon. Such materials include InSb, InGaAs, InAs, InAsSb and Ge. Generally, transistors fabricated with these materials can be operated with a lower supply voltage than MOSFETs. For instance, *quantum well field effect transistors* (*QWFETs*) with InSb channels have been demonstrated with a 10× reduction in dynamic power dissipation and 50× reduction in switching energy compared to advanced silicon MOSFETs [42]. Despite their relatively lower supply voltage, the fabricated devices have a 50% higher intrinsic switching frequency, and a 2.8× lower intrinsic gate delay, thanks to their higher carrier mobility.

1.3.3 Novel Information Processing Technologies

This category of emerging technologies is characterized by a different physics explaining the information processing procedure. While in CMOS and CMOS extensions electrons represent the state variable, which are field-effect controlled in an inversion, depletion or enhancement layer, in this category of emerging technologies, the information may be processed not only as electrons, but also as spin or molecular state and the underlying physics is generally quantized in nature.

1.3.3.1 Single-Electron Transistors

Single-electron transistors (*SET*) are three-terminal devices having a similar organization to MOSFET (Fig. 1.10a); i.e., with one terminal (gate) electrostatically controlling the current flow between the other two terminals (drain and source). However, the channel is replaced by a quantum dot with quantized electron states, which can be occupied by tunneling electrons from drain to source, or vice versa, through tunnel junctions separating the drain and source from the quantum dot. The gate electrostatically controls the quantum states in the quantum dot, thus controlling the current flow between drain and source.

Potentially, SETs can be used as replacements of MOSFETs for Boolean logic operations. However, the need of novel architectures is required in order to address the issue of variability affecting the parasitic capacitances, which highly influence the device behavior. Once such architectures are defined and optimized, SETs will have a large benefit in terms of low power and high integration density. Among these novel architectures, a hybrid SET/MOSFET has been reported as a promising approach for multi-valued logic circuits and memories [43]. Also pattern matching circuits and hardware dedicated to recognition tasks may be efficiently implemented with SETs [44, 45].

1.3.3.2 Molecular Devices

Molecules of interest are two-terminal devices that can store information or perform computation with passive elements, *e.g.*, rotaxanes depicted in Fig. 1.10b. Their physics is generally based on either charge trapping in a similar way to Coulomb blockades, or on a changing resistivity depending on the molecular state. The most interesting use of molecular devices is their application as latching switches. This function results from the hysteresis that enables the operation of the molecular devices as programmable diodes. Such diodes cover several categories of devices: metallic filament formation and dissolution along the molecule [46], charge trapping [47, 48] and change of the molecular configuration [49].

The interest in using molecular devices is their potentially higher integration density, their very low power, and the possibility of leveraging their self-assembly in order to simplify the large-scale fabrication process. However, there is a discrepancy among the published results on the performance of molecular devices and on their technological integration.

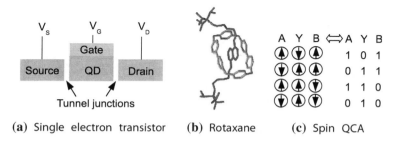

(a) Single electron transistor   (b) Rotaxane   (c) Spin QCA

Fig. 1.10 Schematic view of novel information processing devices: **a** SETs have a quantum dot (QD) with discretized electron states as a channel. **b** Rotaxanes are an example of molecular switches with a ring translating along an axel. **c** Spin QCA realizing $Y = \overline{A \cdot B}$ with spin up $= 1$ and down $= 0$ [54]

The integration of molecular switches into large scale systems is expected to be within hybrid systems with both CMOS and molecular devices. The *CMOS-molecular* (*CMOL*) hybrid circuit concept is based on a linking CMOS circuit parts with crossbars fabricated with silicon nanowires and molecular latching diodes at the crosspoints of the nanowires [50, 51] and will be explained in Sect. 1.4.6. Many circuits were simulated with the CMOL concept including *field-programmable logic arrays* (*FPGA*) [50], crossbar memories [52] and hardware dedicated to pattern classification [53]; and they predict a large improvement in comparison with their implementation in standard CMOS.

1.3.3.3 Spin Devices

The idea of using electron spins as information state derives from the attractive concept of using electron wave functions to transmit information with no material (i.e., electron) transport. In fact, electrons trapped in quantum dots have overlapping wave functions, and can influence the state of each other mutually. However, the high variability and sensitivity to noise makes this approach generally impractical unless the device is operated at low temperature. On the contrary, the spin is less sensitive to noise and parameter variability, it is therefore more attractive to use the coupling of the spin states of electrons in order to process and transmit data [54]. The spin coupling was used in order to implement logic gates such as AND, OR majority voters [54, 55]. The overall organization of spin devices is in arrays of quantum dots in QCAs (Fig. 1.10c). The technology promises an extremely low energy consumption per switching operation. However, it is still sensitive to noise, and spin devices have no gain, meaning that the signal can flow indifferently from input to output or in the opposite direction.

1.4 Regular Architectures and Fabrics

Research on previously surveyed emerging technologies is conducted at the same time as the investigations of potential architectural paradigms for these technologies are carried out. There are several facts that motivate this double approach. On the one hand, the limited abilities of photolithography at the latest technology nodes promotes the use of specific layout patterns. On the other hand, the increasing overall variability at all levels can be addressed by increasing the redundancy and the regularity of the circuit. Also the lack of ability to place some sub-lithographic devices fabricated with emerging technologies may necessitate a solution at the architectural level.

This situation leads to the development of some emerging architectures that share some aspects related to redundancy and regularity. For instance, homogeneous and heterogeneous many-core systems are proposed not only in CMOS technology but also as a hybridized form of CMOS with emerging technologies.

1.4.1 The Need for Regularity

In Sect. 1.2.1 the challenging task of performing photolithography with a wave length that is larger than the critical layout dimensions was explained, and its impact on the design rules was presented. Today, photolithography teams are not able to completely validate a technology process for any design pattern because of the mutual effects on neighboring patterns and the large number of possible patterns within a standard library that create catastrophic proximity effects. In order to simplify the OPC calculation, and avoid the use of non-validated patterns, the solution presented in Sect. 1.2.1 consists in using limited patterns within every basic cell in the library such as logic gates, i.e., only discretized values of critical lines such as *poly-crystalline silicon (poly-Si)* lines are allowed and drawn on a predefined regular grid. It is highly desirable to generalize this approach by defining a macroregularity at the circuit level [12]. The use of microregularity alone as suggested in Sect. 1.2.1 can have some penalties if the patterns at the borders between the cell libraries are not RET-compatible. The concept of macroregularity can be seen in FPGA and memory cells and it may provide considerable benefits in overcoming the penalties of microregular designs. For this reason, FPGA and memories are often the first product to come to market in a new technology process.

The difficulties related to photolithography have an impact on different design levels. For instance, the variability affecting the line width and the discrepancy between drawn and fabricate dimensions affect the basic technology parameters such as channel length and channel doping level. This variability is passed to the device level, causing a large variability in the threshold voltage and off-current for instance; which influence the system design by causing a fluctuation of the delay and power dissipation. The traditional, yet still attractive way to design robust systems with respect to these variability issues, is to make the system redundant by including different instances of the same computational units at different granularity level in order to improve the fault tolerance. This idea was developed in [56, 57] by introducing an approach reminiscent to neural networks, based on a redundant computation whose result is weighted and summed. Then, a thresholder restores the computation output in a more reliable way, even if the single instances of the computational unit are defect-prone.

These motivations towards regular architectures are particular to photolithographic processes of CMOS technology. When it comes to emerging technologies, the motivations are different, but the final circuit architecture may share some aspects with advanced CMOS technology. Section 2.1 will survey the different NW fabrication techniques and explains the fundamental difference between classes of technologies with respect to their ability to control the nanowire placement. Similarly, the growth direction of CNTs is not accurately controlled, so that an accurate placement of single CNTs is impractical unless low-throughput and VLSI-incompatible means are used [58, 59]. Consequently, self-assembly of quasi-one dimensional devices is an attractive way to arrange them in cost- and time-efficient ways [26]. This technique yields parallel layers or crossing arrays of the considered structures and motivates the use of regular architectures with emerging technologies.

1.4.2 CMOS Many-Core Architectures

Regular architectures are today already implemented within commercial products with CMOS technology. Currently, dual-core chips are very common, while quad-core chips are entering the market place, and a demonstration of an 80-core chip has recently been published [60]. The overall trend is towards an FPGA-like organization of many objects, in the so-called *next-generation FPGAs*, such as the *Raw* architecture depicted in Fig. 1.11 [61]. These objects can be similar, for instance a set of identical general purpose processors including the required memory; in this case, the architecture is called homogeneous. If the objects are different, *i.e., application specific integrated circuits (ASIC)* or *systems-on-chip (SoC)*, then the architecture is called heterogeneous.

Such large many-core architectures raise several issues that represent the topics of many research fields. On the one hand, the communication between the different cores can be the bottleneck to be addressed in order to avoid the performance degradation due to the signal delay or any possible congestion. The *network-on-chip (NoC)* paradigm solves this issue in a very efficient way [23] by providing both the hardware to interface the cores and the communication protocol for the network. On the other hand, there is an issue at the software and algorithmic level: given such a large number of cores, it is necessary to optimize the task management by compiling the software properly in order to maximize the performance, improve the yield and reduce the power supply.

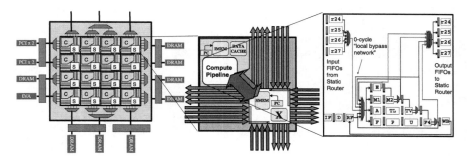

Fig. 1.11 Raw architecture as an example for CMOS many-core regular architecture: it comprises 16 tiles; formed each by a processor, routers, network wires and memories [61]

1.4.3 Via Patterned Gate Arrays

The *via-patterned gate array* (*VPGA*) concept is an emerging architectural paradigm that is motivated by two different facts. On the one hand, the necessity of designing regular structures with discretized line width values following a regular grid, as explained in Sect. 1.2.1, motivates for an overall organization reminiscent to FPGA cells, with an internal regularity of the basic blocks, *complex logic blocks* (*CLB*). On the other hand, the high cost of FPGAs in terms of *energy-delay-product* (*EDP*) is very high compared to ASIC implementations. The VPGA approach is a compromise between ASIC and FPGA solutions with a fully regular design [62, 63]. The novelty of the approach is that a part of the configurable circuit is one time mask-programmed during manufacturing. This part concerns the interconnects and a part of the logic formed by the lowest level of the *look-up table* (*LUT*) tree Fig. 1.12.

1.4.4 Crossbar Architecture

Previously to the emergence of the crossbar architecture, many experiences were performed with a massively parallel computer built in Hewlett-Packard laboratories, the Teramac [64], in state-of-the-art CMOS technology. Despite the high defect rate affecting single components in the Teramac, the approach seemed to be efficient, resulting in 100× faster robust operation than high-end single a processor computer in some configurations. The required architectural elements are a large number of computing instances, a parallelism of their operation and a high bandwidth.

Since these elements naturally exist in the crossbar architecture, this architectural paradigm emerged as a possible approach for reliable massive and parallel computing with highly defective basic components [65]. A crossbar has been conceived as a double layer of parallel nanowires laid out in perpendicular directions. At the nanowire crossings, called crosspoints, molecular devices can perform logic

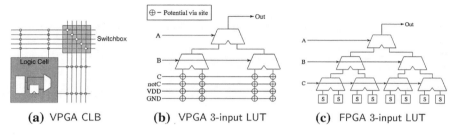

(a) VPGA CLB   **(b)** VPGA 3-input LUT   **(c)** FPGA 3-input LUT

Fig. 1.12 Baseline VPGA architecture [63]: **a** CLB and switch box. **b** VPGA-style LUT: input *C* is configured with vias. **c** FPGA-style LUT: all inputs are configured with software

operation or information storage (Fig. 1.13a). The basic elements of crossbars, i.e. nanowires and molecular switches, are defined at a sub-photolithographic scale with different techniques that will be explained in Sect. 2.2. Due to the immature fabrication technique, a high variability characterizes their electrical properties.

A couple of crossbar prototypes were fabricated with different sizes [36, 66, 67], and the basic function that those prototypes implemented is information storage. Crossbars implementing computational units, such as the *nanoBlock* [68, 69], are also conceptually possible, however they need restoration stages and latches that can be performed by using *resonant tunneling diodes* (*RTDs*) or by hybridizing crossbars with CMOS (Fig. 1.13b). The CMOS part can also provide the necessary gain and input/output interface. It is not excluded that the CMOS part performs more functions than the crossbars in a hybrid architecture, however, the parallelism, reconfigurability and high connectivity will be the main advantages provided by crossbars thanks to their matrix form, in addition to their ability to scale down below the limit imposed by photolithography.

1.4.5 NanoPLA

The *nanoPLA* architecture is a concept based on semiconducting SiNWs organized in a crossbar fashion with molecular switches at their crosspoints. The switches can be programmed in order to perform either signal routing or wired-OR logic function. The input of the crossbar represents a decoder, which is used in order to uniquely address every nanowire independently of the others. The decoder design

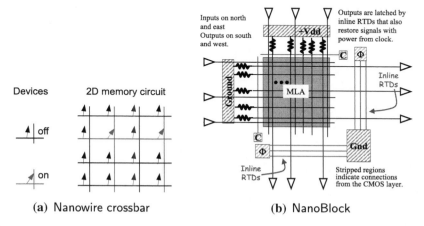

(a) Nanowire crossbar   (b) NanoBlock

Fig. 1.13 Crossbar architecture: **a** Nanowire crossbar with configurable switches at the crosspoints [66]. **b** NanoBlock including a crossbar and an output plane that restores and latches signals with RTDs [69]

assumes that the nanowires are differentiated by a certain doping profile [31]. This will be explained with more details in Sect. 3.2. The output of the crossbar is routed to a second crossbar, in which the signals can be inverted by gating the nanowires carrying the signals. A cascade of these two planes is equivalent to a NOR plane. Two back-to-back NOR planes are universal gates, and they can implement the traditional AND-OR PLA by applying DeMorgan's theorem [70]. A nanoPLA block is illustrated in Fig. 1.14.

1.4.6 CMOS/Molecular Hybrid Systems

The CMOL approach combines CMOS with molecular and one-dimensional devices. The purpose of this architecture is to solve two main issues: first, the alignment and connectivity between the CMOS and the sublithographic part of the circuit in the crossbar; second, the integration of active devices providing gain, since the molecular devices in the crossbar act as passive devices [71].

The CMOL concept has not been implemented in hardware yet. It assumes a certain fabrication process that is feasible with the state-of-the-art fabrication facilities. The basic idea is to define a grid of CMOS lines that are terminated by pointed metallic pins with two different heights. Then, a nanowire crossbar is defined after the backend steps of the CMOS processing. By a sequence of etching and planarization, the two layers of nanowires can be contacted by the CMOS pins depending on their height.

In order to insure the alignment between the nanowires and the CMOS lines defined at two different scales, the crossbar is tilted by a certain angle with respect to the CMOS grid; thus making every nanowire connected by two pins. The idea of bridging the scales by tilting the structures has been proposed for the first time in

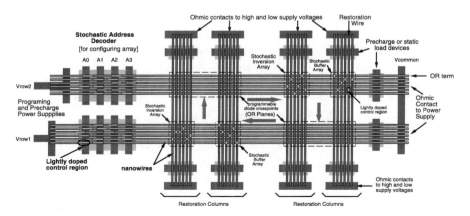

Fig. 1.14 NanoPLA architecture: every nanoPLA block comprises two back-to-back NOR plane. Every plane has two crossbars implementing OR and inversion respectively [70]

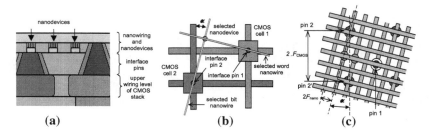

Fig. 1.15 CMOL architecture: **a** Schematic cross-section of CMOS and crossbar parts. **b** Addressing of sublithographic nanowires. **c** Addressing of two crosspoints [71]

the seventies within the CMOS context. The CMOL approach gave this idea a new breath in the context of emerging technologies. The concept is illustrated in Fig. 1.15.

In CMOL circuits, the crossbar part represents a programmable interconnect grid, and it can perform the wired-OR function without any inversion, while the logic functions, including the signal inversion, are performed by the CMOS part. This part is dedicated as well to the input/output interfacing, and the decoding of the nanowire crossbar. Many circuits have been simulated with the CMOL approach, including FPGA-like programmable digital logic architecture [50] and biologically inspired circuits for image recognition [72], and they promise a better performance and more fault-tolerance than stand-alone CMOS.

1.5 Challenges of Regular Emerging Architectures

Among the previously surveyed regular emerging architectures, this section focuses on those involving both the CMOS and the crossbar technology, and it presents a set of challenges that are the core of intensive research efforts. The challenges are numerous, and they concern many levels of design: device fabrication, circuit design, and system level design including CAD tools. In the second part of this section, the focus will be on CNT technology by highlighting its challenges at different levels, and introducing the main problem dealing with CNTs that will be addressed in this work.

1.5.1 SiNW Technology

Despite the maturity of CMOS technology, its hybridization with the crossbar circuits is very challenging from the technological point of view because of the lack of maturity of the crossbar technology. As a matter of fact, some fabrication techniques of silicon nanowires, especially the bottom-up approaches as it will be explained in Sect. 2.1.1, undergo a high variability with respect to the nanowire dimensions (cross-section, length), lattice structure, surface states, etc. The process needs to be controlled in a more accurate way in order to guarantee the uniformity of the electrical properties of the nanowires. It is also desirable to improve the ability to design and control the synthesis of the molecular switches in order to enhance their electrical properties, namely their switching speed, and their on- and off-resistance.

In addition to the process variability affecting both the nanowires and the molecular switches, the integration of these elements into three layers (a layer of molecular switches sandwiched between two nanowire layers) is still a challenging issue. It is necessary to improve the deposition techniques in order to avoid the performance degradation of the molecular switches after the top nanowire layer is deposited. Also the linearity of nanowire layers shows a large discrepancy, depending on the process used to align nanowires. On the other hand, a hybrid process including both crossbars and CMOS technology needs a special attention to the temperature budget: the molecular switches have an organic part and they are generally processed at temperatures lower than 300°C. Thus, the crossbar part of the circuit has to be processed with the backend steps, or at least between the front- and the backend steps.

The design of such hybrid circuits has its specific challenges as well. Unlike stand-alone CMOS circuits that are defined on the photolithographic scale, hybrid emerging circuits have the crossbar part that is designed on a sub-photolithographic scale. The difference between these two scales can be around $5\times$ to $10\times$, depending on the nanowire dimensions and the available photolithography. One of the challenging tasks is to bridge these two scales by using a decoder, thus allowing every nanowire to be uniquely addressed and controlled by the CMOS part of the chip. Besides the decoder problem, the design of universal gates with the crossbar part in not an easy task. Even though wired-OR functions can be implemented in a straightforward way with passive molecular switches operating as diodes, the implementation of inversion necessitates the utilization of devices with gain. The field-effect control of nanowires in a crossbar is possible in general, but performing this in a sublithographic array, is a very challenging task.

The design using regular emerging technologies requires methodologies that focus on the inherently high defect rate of single components. It is therefore necessary to develop defect models and to privilege designs that are defect-tolerant. The design of self-checking testing units is also fundamental, since a large part of the sub-photolithographic circuit is expected to be defective. Another challenging aspect is the design of re-configurable circuits, whose implemented function can vary from chip to chip depending on the defect distribution, and it can also vary within the same chip during its life cycle according to the changing defect density with the component usage. In addition to yield, performance is also a fundamental aspect that designers have to consider carefully. It is possible that hybrid circuits do not outperform their CMOS counterparts, because of the possibly larger delay of crossbars due to the high nanowire resistivity and capacitance. However, their scaling ability and reconfigurability may be their most important assets. It is therefore important to design parallel circuits at not too large granularity levels, in order to leverage the potential of parallel and re-configurable computing.

The system level design will have to cope with these problems by developing the necessary design tools and methodologies that take into account the fundamental differences between hybrid crossbar/CMOS circuits and bare CMOS circuits. It is therefore highly desirable to address the problem of mapping logic functions onto defective crossbars. Instead of replacing a defective crossbar by a defect-free one, it is possible to map a part of the logic function onto the safe part of the defective crossbar, provided that the technology mapping tool is aware of the map of physical defects and it uses it during the function is mapped onto the hardware. The logic synthesis can be also made aware of the defect-distribution in the crossbar, in such a way that the synthesized circuits have an certain level of redundancy, for instance by covering the logic functions more than once.

1.5.2 CNT Technology

Carbon nanotube technology has emerged less than nanowire technology in regular circuits, mainly because of technological reasons. Even though CNTs promise a higher performance than bulk Si and SiNW technologies, the hard control of CNT growth (chirality, diameter) and doping make research focus more on the technological challenges. Thus, a full system integration with CMOS has not been consistently suggested yet, even though a few preliminary tries have been reported [73]. Regularity seems to be a natural choice in CNT technology, because CNT devices have to be aligned on the substrate in large parallel layers for a better electrical performance. Changing the CNT orientation arbitrarily like MOSFET channels is technologically difficult to realize, since CNTs are transferred to the substrate to be functionalized simultaneously. It is technically thinkable though, to define CNT transistors with different channel orientations, if some area penalty is accepted. Design with regular CNT devices and logic gates will be exploited in the last chapter of this book.

As discussed above for regular circuits based on SiNW crossbars, the challenges of CNT technology are diverse. Despite the technological difficulties mentioned previously for CNTs, the design with these structures needs to take into account the high variability, and the difficulty of carrying out an n-type chemical doping. In the design approach that will be proposed in this work, the question is about leveraging some intrinsic properties of CNT technology, that make it fundamentally different from bulk Si technology. For instance, the use of intrinsic CNTs, motivated by the challenges of the chemical doping, represents an interesting opportunity for designers to explore the options offered by the ability to electrically control the CNT "doping".

Many other challenges exist at the system level, but they will not be addressed in this work. For instance, the creation of design tools and methods that yield a robust logic cell design against the variability of the CNT alignment. Also the variability affecting the CNT chirality, resulting in a badly controllable metallic and semiconducting behavior of the fabricated devices, can be addressed by a precise modeling of the variability and by taking this model into account while designing the logic circuits. It is consequently possible to enhance the design flow with some inputs related to the variability level, in order improve the fault-tolerance of the synthesized system.

1.6 Organization of the Book

The book is organized in four main chapters following the introduction and a concluding chapter. In the next three chapters, the focus is on crossbar technology from fabrication to system design. Then, the following chapter focuses on design aspects for CNT technology.

In Chap. 2, the technology part of this work is presented. This part includes a literature survey on nanowire fabrication techniques, as well as different technologies which have been successfully reported for the fabrication of crossbar circuits. Then, the available facilities at EPFL are presented, while the technological choices made with respect to this infrastructure are explained. Then, the fabrication process and the steps undertaken to optimize the process are presented. Thereafter, the results are demonstrated by means of structural and electrical characterization of the fabricated devices. Possible applications of the technology are proposed in the following section. Finally, a discussion of the novelty of this part and a summary of its contributions are given.

In Chap. 3, the focus is on the design aspect related to the introduced crossbar technology. This chapter mainly deals with the optimal encoding scheme for nanowire crossbars. After an introduction recalling the baseline crossbar architecture, a survey of different decoders reported in literature for nanowire crossbars is presented. Then, the multi-valued encoding approach is introduced and it is evaluated in terms of yield and area. Thereafter, the concept of a specific decoder for the fabrication technique given in Chap. 2 is presented and evaluated in terms of yield, area and fabrication complexity. This chapter is concluded with a discussion of this part of the work and a summary of its contributions.

In Chap. 4, the system level design aspects following from the problem of encoding addressed in Chap. 3 are investigated. In this part of the book, a current-based test is proposed, which maximizes the ability of detecting badly addressed nanowires. This chapter starts by introducing the need for this testing procedure and how it can be performed. Then, it presents the error model assumed for badly addressed nanowires. Thereafter, the model is implemented; and simulation results illustrate its benefits. Finally, a discussion of this part of the work and a summary of its contributions conclude the chapter.

In Chap. 5 the considered technology is based on CNTs. This chapter focuses on novel design techniques to be applied on ambipolar CNT devices. It starts by surveying the utilization of CNTs to fabricate transistors and logic gates, and the literature related to ambipolar CNTs. Then, a dynamic and a static logic design approach are introduced, and the synthesis of multi-level logic circuits with static logic is demonstrated. Thereafter, various approaches to design regular fabrics with ambipolar CNTs are explained. The chapter is concluded with a discussion of this part of the work and a summary of its contributions.

In Chap. 6, the work is concluded and possible future directions are presented.

References

1. International technology roadmap for semiconductors (ITRS) (2007) http://www.itrs.net/reports.html. Tech. Rep., 2007
2. Cerofolini G (2007) Realistic limits to computation. II. The technological side. Appl Phys A 86(1):31–42
3. Lilienfeld JE (1925) Method and apparatus for controlling electric current. US Patent No. 1745175
4. Bardeen J, Brattain WH (1948) Three-electrode circuit element utilizing semiconductor materials. US Patent No. 2524035
5. Shockley W (1948) Circuit element utilizing semiconductive material. US Patent No. 2569347
6. Kahng D (1960) Electric field controlled semiconductor device. US Patent No. 2524035
7. Moore GE (1965) Cramming more components onto integrated circuits. Electron Mag 38:114–117
8. Skotnicki T (2006) CMOS technologies for end of roadmap. Course at Ecole Polytechnique Fèdérale de Lausanne, Switzerland
9. Dennard RH, Gaensslen FH, Yu HN, Rideout VL, Bassous E, Leblanc AR (1974) Design of ion-implanted MOSFET's with very small physical dimensions. IEEE J Solid-State Circuits 9(5):256–268
10. Wang J, Wong AKK, Lam EY (2004) Performance optimization for gridded-layout standard cells. Proc SPIE 5567(1):107–118
11. Lavin M, Heng FL, Northrop G (2004) Backend CAD flows for restrictive design rules. In: Proceedings of the International Conference on Computer-Aided Design, pp 739–746
12. Calhoun B, Cao Y, Li X, Mai K, Pileggi L, Rutenbar R, Shepard K (2008) Digital circuit design challenges and opportunities in the era of nanoscale CMOS. Proc IEEE 96(2):343–365
13. Liebmann LW, Barish AE, Baum Z, Bonges HA, Bukofsky SJ, Fonseca CA, Halle SD, Northrop GA, Runyon SL, Sigal L (2004) High-performance circuit design for the RET-enabled 65-nm technology node. In: Proceedings of SPIE Design and Process Integration for Microelectronics Manufacturing II, vol 5379, no 1, pp 20–29
14. Masetti G, Severi M, Solmi S (1983) Modeling of carrier mobility against carrier concentration in arsenic-, phosphorus-, and boron-doped silicon. IEEE Trans Electron Devices 30(7):764–769
15. Lombardi C, Manzini S, Saporito A, Vanzi M (1988) A physically based mobility model for numerical simulation of nonplanar devices. IEEE Trans Computer-Aided Des Integr Circuits Syst 7(11):1164–1171
16. Ernst T, Andrieu F, Weber O, Duprt C, Faynot O, Ducroquet F, Clavelier L, Hartmann J, Barraud S, Ghibaudo G, Deleonibus S (2006) High-mobility nano-scaled CMOS: some opportunities and challenges. In: 1990 Symposium of the VLSI Technology, pp 1–2
17. Hoffmann T, Doorribos G, Ferain I, Collaert N, Zimmerman P, Goodwin M, Rooyackers R, Kottantharayil A, Yim Y, Dixit A, De Meyer K, Jurczak M, Biesemans S (2005) GIDL

(gate-induced drain leakage) and parasitic schottky barrier leakage elimination in aggressively scaled HfO2/TiN FinFET devices. In: Technical Digest—International Electron Devices Meeting, IEDM, pp 725–728

18. Hori T (1990) Drain-structure design for reduced band-to-band and band-todefect tunneling leakage. pp 69–70

19. Watanabe H, Matsuzawa K, Takagi S (2003) Scaling effects on gate leakage current. IEEE Trans Electron Devices 50(8):1779–1784

20. Pott V (2008) Gate-all-around silicon nanowires for hybrid single electron transistor/CMOS applications. PhD dissertation, Lausanne. http://library.ep.ch/theses/?nr=3983

21. Osburn CM, Bellur KR (1998) Low parasitic resistance contacts for scaled ULSI devices. Thin Solid Films 332(1–2):428–436

22. Asenov A, Brown AR, Davies JH, Kaya S, Slavcheva G (2003) Simulation of intrinsic parameter uctuations in decananometer and nanometer-scale MOSFETs. IEEE Trans Electron Devices 50:1837–1852

23. Jalabert A, Murali S, Benini L, De Micheli G (2008) Design, automation, and test in Europe: the most in uential papers of 10 years DATE. Springer, Heidelberg

24. Arnold MS, Green AA, Hulvat JF, Stupp SI, Hersam MC (2006) Sorting carbon nanotubes by electronic structure using density differentiation. Nat Nanotechnol 1:60–65

25. Zhang G, Qi P, Wang X, Lu Y, Li X, Tu R, Bangsaruntip S, Mann D, Zhang L, Dai H (2006) Selective etching of metallic carbon nanotubes by gas-phase reaction. Science 314(5801):974–977

26. Patil N, Lin A, Myers E, Wong HS, Mitra S (2008) Integrated wafer-scale growth and transfer of directional carbon nanotubes and misaligned-carbon-nanotube-immune logic structures. In: 2008 Symposium of the VLSI Technology, pp 205–206

27. Liu J, O'Connor I, Navarro D, Gaffiot F (2007) Novel CNTFET-based reconfigurable logic gate design. In: Annual ACM IEEE Design Automation Conference, pp 276–277

28. Lin YM, Appenzeller J, Avouris P (2004) Novel carbon nanotube FET design with tunable polarity. In: IEEE International Electron Devices Meeting 2004. IEDM Technical Digest, pp 687–690

29. Lin YM, Appenzeller J, Knoch J, Avouris P (2005) High-performance carbon nanotube field-effect transistor with tunable polarities. IEEE Trans Nanotechnol 4(5):481–489

30. Kang SJ et al (2007) High-performance electronics using dense, perfectly aligned arrays of single-walled carbon nanotubes. Nat Nanotechnol 2(4):230–236

31. Gudiksen MS, Lauhon LJ, Wang J, Smith DC, Lieber CM (2002) Growth of nanowire superlattice structures for nanoscale photonics and electronics. Nature 415:617–620

32. Yang C, Zhong Z, Lieber CM (2005) Encoding electronic properties by synthesis of axial modulation-doped silicon nanowires. Science 310(5752):1304–1307

33. Lauhon LJ, Gudiksen MS, Wang D, Lieber CM (2002) Epitaxial core-shell and core-multishell nanowire heterostructures. Nature 420:57–61

34. Huang Y, Duan X, Wei Q, Lieber CM (2001) Directed assembly of one-dimensional nanostructures into functional networks. Science 291(5504):630–633

35. Chen Y, Ohlberg DAA, Li X, Stewart DR, Stanley Williams R, Jeppesen JO, Nielsen KA, Stoddart JF, Olynick DL, Anderson E (2003) Nanoscale molecular-switch devices fabricated by imprint lithography. Appl Phys Lett 82:1610–1612

36. Wu W, Jung GY, Olynick DL, Straznicky J, Li Z, Li X, Ohlberg DAA, Chen Y, Wang SY, Liddle JA, Tong WM, Williams RS (2005) One-kilobit cross-bar molecular memory circuits at 30-nm half-pitch fabricated by nanoimprint lithography. Appl Phys A Mater Sci Process 80(6):1173–1178

37. Melosh NA, Boukai A, Diana F, Gerardot B, Badolato A, Petroff PM, Heath JR (2003) Ultrahigh-density nanowire lattices and circuits. Science 300(5616):112–115

38. Sacchetto D, Ben-Jamaa MH, De Micheli G, Leblebici Y (2009) Fabrication and characterization of vertically stacked gate-all-around Si nanowire FET arrays. In: ESSDERC 2009

39. Doherty L, Liu H, Milanovic V (2003) Application of MEMS technologies to nanodevices. In: ISCAS'03. Proceedings of the 2003 International Symposium on Circuits and systems, vol 3, pp III–934–III–937
40. Ng RMY, Wang T, Chan M (2007) A new approach to fabricate vertically stacked single-crystalline silicon nanowires. pp 133–136
41. Moselund KE, Bouvet D, Ben Jamaa HH, Atienza D, Leblebici Y, De Micheli G, Ionescu AM (2008) Prospects for logic-on-a-wire. Microelectron Eng 85:1406–1409
42. Datta S (2007) III-V field-effect transistors for low power digital logic applications. Microelectron Eng 84(9–10):2133–2137
43. Inokawa H, Fujiwara A, Takahashi Y (2003) A multiple-valued logic and memory with combined single-electron and metal-oxide-semiconductor transistors. IEEE Trans Electron Devices 50(2):462–470
44. Saitoh M, Harata H, Hiramoto T (2004) Room-temperature demonstration of integrated silicon single-electron transistor circuits for current switching and analog pattern matching. pp 187–190
45. Bandyopadhyay S, Roychowdhury V (1996) Computational paradigms in nanoelectronics: quantum-coupled single electron logic and neuromorphic networks. Jpn J Appl Phys 35:3350–3362
46. Stewart DR, Ohlberg DAA, Beck PA, Chen Y, Williams RS, Jeppesen JO, Nielsen KA, Stoddart JF (2004) Moleculeindependent electrical switching in Pt/organic monolayer/Ti devices. Nano Lett 4(1):133–136
47. Rotenberg E, Venkatesan R (2006) The state of ZettaRAM. In: 1st IEEE International Conference on Nano-Networks, pp 1–5
48. Akkerman HB, Blom PWM, de Leeuw DM, de Boer B (2006) Towards molecular electronics with large-area molecular junctions. Nature 441:69–72
49. Dichtel WR, Heath JR, Fraser Stoddart J (2007) Designing bistable [2]rotaxanes for molecular electronic devices. Royal Soc Lond Philos Trans Ser A 365:1607–1625
50. Strukov DB, Likharev KK (2005) CMOL FPGA: a reconfigurable architecture for hybrid digital circuits with two-terminal nanodevices. Nanotechnology 16:888–900
51. Snider GS, Williams RS (2007) Nano/CMOS architectures using a field-programmable nanowire interconnect. Nanotechnology 18(3):035 204–035 215
52. Strukov DB, Likharev KK (2007) Defect-tolerant architectures for nanoelectronic crossbar memories. J Nanosci Nanotechnol 7(1):151–167
53. Lee JH, Likharev KK (2007) Defect-tolerant nanoelectronic pattern classifiers. Int J Circuit Theory Appl 35(3):239–264
54. Bandyopadhyay S, Das B, Miller AE (1994) Supercomputing with spin-polarized single electrons in a quantum coupled architecture. Nanotechnology 5:113–133
55. Imre A, Csaba G, Ji L, Orlov A, Bernstein GH, Porod W (2006) Majority logic gate for magnetic quantum-dot cellular automata. Science 311:205–208
56. Schmid A, Leblebici Y (2004) Robust circuit and system design methodologies for nanometer-scale devices and single-electron transistors. IEEE Trans Very Larg Scale Integr (VLSI) Syst 12(11):1156–1166
57. Schmid A, Leblebici Y (2003) A modular approach for reliable nanoelectronic and very-deep submicron circuit design based on analog neural network principles. vol 2, pp 647–650
58. Nagahara LA, Amlani I, Lewenstein J, Tsui RK (2002) Directed placement of suspended carbon nanotubes for nanometer-scale assembly. Appl Phys Lett 80(20):3826–3828
59. Williams PA, Papadakis SJ, Falvo MR, Patel AM, Sinclair M, Seeger A, Helser A, Taylor RM II, Washburn S, Superfine R (2002) Controlled placement of an individual carbon nanotube onto a microelectromechanical structure. Appl Phys Lett 80(14):2574–2576
60. Vangal S, Howard J, Ruhl G, Dighe S, Wilson H, Tschanz J, Finan D, Iyer P, Singh A, Jacob T, Jain S, Venkataraman S, Hoskote Y, Borkar N (2007) An 80-tile 1.28-TFLOPS network-on-chip in 65nm CMOS. In: 1st IEEE International Conference on Nano-Networks, pp 98–589
61. Taylor M, Psota J, Saraf A, Shnidman N, Strumpen V, Frank M, Amarasinghe S, Agarwal A, Lee W, Miller J, Wentzlaff D, Bratt I, Greenwald B, Hoffmann H, Johnson P, Kim J (2004)

Evaluation of the Raw microprocessor: an exposed-wire-delay architecture for ILP and streams. In: Proceedings of the 31st Annual International Symposium on Computer Architecture, pp 2–13

62. Tong K, Kheterpal V, Rovner V, Pileggi L, Schmit H (2003) Regular logic fabrics for a via patterned gate array (VPGA). In: Proceeding of the Custom. Integrated Circuits Conference, pp 53–56

63. Pileggi L, Schmit H, Strojwas AJ, Gopalakrishnan P, Kheterpal V, Koorapaty A, Patel C, Rovner V, Tong KY (2003) Exploring regular fabrics to optimize the performance-cost trade-off. In: DAC'03: Proceedings of the 40th conference on Design automation, pp 782–787

64. Culbertson W, Amerson R, Carter R, Kuekes P, Snider G (1997) Defect tolerance on the Teramac custom computer. In: Proceedings of the 1997 IEEE Symposium on FPGA's for Custom Computing Machines, pp 116–123

65. Heath JR, Kuekes PJ, Snider GS, Williams RS (1998) A Defecttolerant computer architecture: opportunities for nanotechnology. Science 280(5370):1716–1721

66. Luo Y, Collier CP, Jeppesen JO, Nielsen KA, DeIonno E, Ho G, Perkins J, Tseng HR, Yamamoto T, Stoddart JF, Heath JR (2002) Two-dimensional molecular electronics circuits. J Chem Phys Phys Chem 3:519–525

67. Green JE, Wook Choi R, Boukai A, Bunimovich Y, Johnston- Halperin E, Deionno E, Luo Y, Sheriff BA, Xu K, Shik Shin Y, Tseng R, Stoddart JF, Heath JR (2007) A 160-kilobit molecular electronic memory patterned at 10^{11} bits per square centimetre. Nature 445:414–417

68. Goldstein SC, Budiu M (2001) NanoFabrics: spatial computing using molecular electronics. In: Proceedings of the 28th Annual International Symposium on Computer Architecture, pp 178–189

69. Goldstein S, Rosewater D (2002) Digital logic using molecular electronics. vol 1, pp 204–459

70. DeHon A, Likharev KK (2005) Hybrid CMOS/nanoelectronic digital circuits: devices, architectures, and design automation. In: ICCAD' 05: Proceedings of the 2005 IEEE/ACM International conference on Computer-aided design, pp 375–382

71. Likharev KK (2007) Hybrid semiconductor/nanoelectronic circuits: freeing advanced lithography from the alignment accuracy burden. J Vac Sci Technol B Microelectron Nanometer Struct 25:2531–2536

72. Tüurel O, Lee JH, Ma X, Likharev KK (2004) Neuromorphic architectures for nanoelectronic circuits. Int J Circuit Theory Appl 32(5):277–302

73. Close GF, Yasuda S, Paul B, Fujita S, Wong HSP (2008) A 1 GHz integrated circuit with carbon nanotube interconnects and silicon transistors. Nano Lett 8(2):706–709

Chapter 2
Fabrication of Nanowire Crossbars

Nanowire crossbars gained an increasing interest in the last years, namely because the fabrication techniques have become more mature and versatile. Parallel research works have been carried out on different levels ranging from device to circuit and system levels in order to identify and address the challenges facing the utilization of this emerging paradigm in the future.

Interestingly, the circuit and system level problems and their proposed solutions depend on some properties of the fabrication techniques. Thus, understanding the fabrication techniques and the device properties enable a better assessment of the global problem. Numerous fabrication technologies have been suggested previously, focusing on different objectives. Some of them focus on the use of CMOS techniques while shrinking the dimensions below the CMOS limits [1]. Other techniques focus on reducing the pitch far below the CMOS limits by using novel methods that are not CMOS-compatible and that may be too expensive for mass production of VLSI systems [2]. In another set of techniques, the focus is to integrate the nanowires into vertical stacks in a CMOS-compatible way, without any optimization of the nanowire pitch [3–5]. These are just example of the different objectives of some previous approaches. A more detailed survey of the available techniques is given in this chapter.

None of the proposed solutions targeted the fabrication of nanowires with sub-lithographic dimensions and pitch, while using only CMOS processing steps. The goal of the work reported in this chapter is to shrink both nanowire dimensions and pitch below the photolithography limit, while keeping the fabrication technique cost-efficient and CMOS-compatible, in the sense that it uses only standard CMOS processing steps. Since the constraint on the technology process is relaxed (i.e., any standard CMOS steps are suitable), then the fabrication is carried out with the available facilities at the *Center for Micro- and Nanotechnologies* (*CMI*) at EPFL [6]. The ambitious fabrication technique that has been developed, has the advantage of using only standard photolithography and CMOS fabrication steps, with the available photolithography limit of 0.8 μm, yet achieving sub-photolithographic device pitch down to less than 40 nm.

M. H. Ben Jamaa, *Regular Nanofabrics in Emerging Technologies*,
Lecture Notes in Electrical Engineering, 82, DOI: 10.1007/978-94-007-0650-7_2,
© Springer Science+Business Media B.V. 2011

Parts of this chapter have been published in [7]. It is organized as follows. First the different nanowire fabrication technologies and the various integration techniques of nanowires into crossbar structures are surveyed. Then, the proposed process flow and the available fabrication facilities at the CMI are presented. Next, the encountered fabrication challenges are discussed and the accordingly optimized process is explained. Thereafter, the structural and electrical characterization results are presented. Finally, the advantages and limits of the developed fabrication technique are discussed and the chapter is concluded with the a summery of the contributions of the work at this level.

2.1 Nanowire Fabrication Techniques

Various nanowire fabrication techniques have been proposed in the last decade. They follow two main paradigms: the so-called bottom-up and top-down approaches. Bottom-up approaches are based on the growth of nanowires by generally using nanoscale metallic catalysts; they are subsequently dispersed into a solution and transferred onto the substrate to be functionalized. In contrast, top-down approaches use various types of patterning technique directly on the functional substrate, which are possibly combined with smart processing methods that will be explained in this section in order to reduce the nanowire thickness below the photolithography limit.

2.1.1 Bottom-up Techniques

2.1.1.1 Vapor-Liquid-Solid Growth

One of the widely used bottom-up techniques is the *vapor-liquid-solid* (*VLS*) process, in which the generally very slow adsorption of a silicon-containing gas phase onto a solid surface is accelerated by introducing a catalytic liquid alloy phase. The latter can rapidly adsorb vapor to a supersaturated level; then the crystal growth occurs from the nucleated catalytic seed at the metal-solid interface. Crystal growth with this technique was established in the 1960s [8] and silicon nanowire growth is today mastered with the same technique.

The VLS process allows for the control of the nanowire diameter and direction growth by optimizing the size and composition of the catalytic seeds and the growth conditions, including temperature, pressure and gas composition in the chamber. In [9], defect-free silicon nanowires were grown in a solvent heated and pressurized above its critical point, using alkanethiol-coated gold monocrystals. Figure 2.1 depicts the growth process and *transmission electron microscope* (*TEM*) images of the grown nanowires. The nanowire diameters were ranging from 40 to 50 Å, their length was about several micrometers, and their crystal orientation was controlled with the reaction pressure.

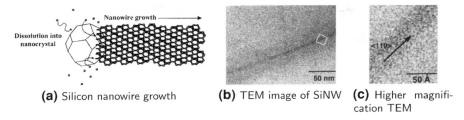

(a) Silicon nanowire growth   **(b)** TEM image of SiNW   **(c)** Higher magnification TEM

Fig. 2.1 Vapor-liquid-solid growth of a silicon nanowire [9]: **a** Free Si atoms from silane dissolve in the Au seed until reaching the Si:Au supersaturation. Then Si is expelled as nanowire. **b** TEM image of SiNW synthesized at 500°C in hexane at 200 bar. **c** TEM of a part of SiNW inside the square in **b** shows high crystalline SiNWs

2.1.1.2 Laser-Assisted Catalytic Growth

A related technique to VLS is the laser-assisted catalytic growth. High-powered, short laser pulses irradiate a substrate of the material to be used for the nanowire growth. The irradiated material either evaporates, sublimates or converts into plasma. Then, the particles are transferred onto the substrate containing the catalyst, where they can nucleate and grow into nanowires. This technique is useful for nanowire materials that have a high melting point, since the laser pulses locally heat the substrate generating the particle for the nanowire growth. It is also suitable for multi-component nanowires, including doped nanowires, and for nanowires with a high-quality crystalline structure [10].

2.1.1.3 Chemical Vapor Deposition

The *chemical vapor deposition* (*CVD*) method was shown to be an interesting technique used with materials that can be evaporated at moderate temperatures [11]. In a typical CVD process, the substrate is exposed to volatile precursors, which react on the substrate surface producing the desired nanowires. In [12], the CVD method was applied to fabricate nanowires based on different materials or combinations of materials, including Si, SiO_2 and Ge.

2.1.1.4 Opportunities and Challenges of Bottom-up Approaches

The bottom-up techniques offer the ability of doping the as-grown nanowires in situ, i.e., during the growth process. In [10], the laser catalytic growth was used in order to control the boron and phosphorus doping during the vapor phase growth of silicon nanowires. The nanowire could be made heavily doped in order to approach a metallic regime, while insuring a structural and electronic uniformity. Another more advanced option offered by the bottom-up approaches consists in alternating the doping regions or the gown materials along the nanowire axis, as

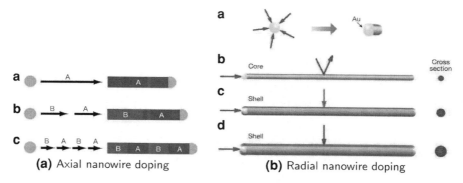

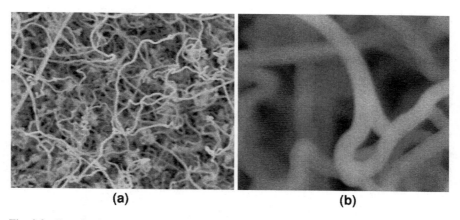

Fig. 2.2 In-situ nanowire doping: **a** Doping along the nanowire axis (axial doping) [13]. **b** Doping around the nanowire axis (radial doping) [12]

Fig. 2.3 Growth of meshed nanowires [14]: **a** SEM image of gold-catalyzed growth of SiNWs on Si_3N_4/Si substrate. Image width = 7 μm. **b** High-magnification image of branched nanowires. Image width = 0.7 μm

illustrated in Fig. 2.2a [13, 15]. The growth of concentric shells with different materials around the nanowire axes was also demonstrated in [12], as illustrated in Fig. 2.2b.

The grown nanowires can either represent a random mesh laid out laterally over the substrate (Fig. 2.3) [14], or they can stand vertically aligned with respect to the substrate (Fig. 2.4) [16, 17]. The growth substrate is in general different from the functional substrate. Consequently, it is necessary to disperse the as-grown nanowires in a solution, and then to transfer them onto the functional substrate, making the process more complex. In [18], the nanowires were dispersed in ethanol; then the diluted nanowire suspension was used to flow-align the nanowires by using microfluidic channels. A similar technique was used in [19] in order to assemble arrays of nanowires through fluidic channel structures formed between

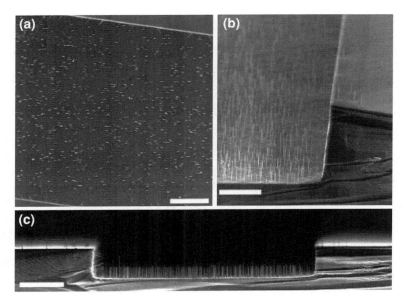

Fig. 2.4 Growth of vertical nanowires [16]: **a** Conformal growth of nanowires to the substrate. **b** Tilted SEM image and **c** a cross-sectional SEM image of the structure. *Scale bars* 10 μm

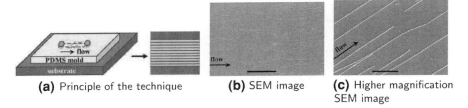

Fig. 2.5 *PDMS*-mold-based assembly of InP nanowires [19]: **a** Schematic representation of the technique. **b** SEM image of the aligned nanowires (*scale bar* = 50 μm). **c** Higher magnification SEM image of the aligned nanowires (*scale bar* = 2 μm)

a *polydimethylsiloxane* (*PDMS*) mold and a flat substrate. This technique yields parallel nanowires over long distances, as shown in Fig. 2.5.

2.1.2 Top-Down Techniques

The top-down fabrication approaches have in common the utilization of CMOS steps or hybrid steps that can be integrated into a CMOS process, while keeping the process complexity low and the yield high enough. They also have in common the ability of defining the functional structures (nanowires) directly onto the functional substrate, with no need of dispersion and transfer of nanowires. Any top-down process uses patterning in a certain way: the patterning technique can be based on a mask, such as in standard photolithography or in other miscellaneous mask-based techniques, or it can be maskless, i.e., using a nanomold for instance.

2.1.2.1 Standard Photolithography Techniques

These techniques use standard photolithography to define the position of the nanowire. Then, by using smart processing techniques, including the accurate control of the etching, oxidation and deposition of materials, it is possible to scale the dimensions down far below the photolithographic limit.

In [20], silicon nanowires were defined on bulk substrates by using CMOS processing steps. First, a Si_3N_4 nitride rib was defined on the substrate. Then, the isotropic etch defined the nanowire underneath the rib. Well controlled self-limited oxidation and subsequent etching steps resulted in silicon nanowires with different cross-section shapes and dimensions, and with a nanowire diameter down to 5 nm.

A related fabrication approach was presented in [21], whereby the nanowire dimensions were defined by an accurate control of the silicon oxidation and etch. The authors transferred the as-fabricated nanowires onto a different substrate in order to arrange them into parallel arrays, which makes this approach partly reminiscent of the bottom-up techniques explained previously.

Another approach was presented in [22], which uses epitaxial Si and Ge layers on a bulk substrate. A thin epitaxial Si layer was sandwiched between Si_3N_4 and SiGe layers. Then, the Si_3N_4 and SiGe were selectively etched, leading to a partial etch of the sandwiched Si layer. The remaining edges of the Si layer were thinned out and lead to 10 nm nanowire diameter.

Using standard photolithography techniques, it could be demonstrated that 3-dimensional vertical stacks of nanowires can be achieved. A possible method was presented in [3, 4], which is based on the alternation of etching and passivation steps in a similar way to the *deep reactive ion etching (DRIE)* technique. This method yields scalloping edges, that can be thinned out through self-limited oxidation and controlled wet etch, resulting in vertical stacks of suspended nanowires. A fully different technique [5] uses alternating epitaxial Si and Ge layers; then, a selective etching of the Ge layers releases the thin suspended Si layers, which can be transformed into suspended Si nanowire stacks by accurately controlling their lateral etch.

2.1.2.2 Miscellaneous Mask-Based Techniques

Instead of using standard lithography and thinning out the devices by means of well controlled oxidation and selective etching, an alternative approach is to use electron-beam lithography [23] that offers a higher resolution below 20 nm, and then eventually further reduce the nanowire diameter by stress-limited oxidation [24].

A higher resolution can be achieved by using *extreme ultraviolet interference lithography (EUV-IL)* [2]. Metallic nanowires with the width of 8–70 nm and a pitch of 50–100 nm could be achieved with this technique. However, this approach needs a highly sophisticated setup in order to provide the required EUV wave length, which is not available in state-of-the-art semiconductor fabrication lines. And it has not be proven so far how this technique may be used in order to fabricate semiconducting nanowires.

The stencil lithography is another approach, which is inherently different from the previous ones, but it shares the same feature of using a mask while avoiding the classical paradigm of CMOS processing, which consists in patterning a photoresist through the mask and then patterning the active layer through the patterned photoresist. The stencil approach [25] is based on the definition of a mask that is fully open at the patterned locations. The mask is subsequently clamped onto the substrate, and the material to be deposited is evaporated or sputtered through the mask openings onto the substrate. Nanowires with a width of 70 nm could be achieved this way. Even though only metallic nanowires have been demonstrated, the technique can be extended to semiconducting nanowires as well.

2.1.2.3 Spacer Techniques

The spacer technique is based on the idea of transforming thin lateral dimensions, in the range of 10–100 nm, into vertical dimension by means of anisotropic etch of the deposited materials. In [26], spacers with a thickness of 40 nm were demonstrated with a line-width roughness of 4 nm and a low variation across the wafer.

In [27], spacers were defined by means of *low pressure chemical vapor deposition (LPCVD)*, then their number was duplicated by using the spacers themselves as sacrificial layers for the following spacer set. This technique, the *iterative spacer technique (IST)*, yields silicon structures with sub-10 nm width and a narrower half-pitch than the photolithography limit.

In [28], the *multi-spacer patterning technique (MSPT)* was developed as in the previous approach, by iterating single spacer definition steps. The spacers were reminiscent to nanowires with a thickness down to 35 nm. The multi-spacer array was not used as a nanomold to define the nanowires, but it was rather used as the actual nanowire layer.

2.1.2.4 Nanomold-Based Techniques

Alternative techniques use the *nanoimprint lithography (NIL)*, which is based on a mold with nanoscale features [29] that is pressed onto a resist-covered substrate in order to pattern it. The substrate surface is scanned by the nanomold in a stepper fashion. The as-patterned polymer resist is processed in a similar way to photolithographically patterned photoresist films. The advantage of this technique is its ability to use a single densely patterned nanomold to pattern a large number of wafers. The obtained density of features on the substrate depends on the density of features in the nanomold, i.e., it mainly depend on the technology used to fabricate the nanomold.

A related technique to NIL, called the *superlattice nanowire pattern transfer technique (SNAP)*, was presented in [30], in which the nanowires are directly defined on the mold; then, they are transferred onto the polymer resist. Nanowires with a pitch of 34 nm could be achieved using the SNAP technique. The superlattice was fabricated by defining 300 successive epitaxial $GaAs/Al_xGa_{1-x}As$ layers on a GaAs wafer. Then, the wafer was cleaved, and the GaAs layers were selectively etched, so that the edge of each layer became an initial nanowire template. Then, a metal was deposited onto the exposed $Al_xGa_{1-x}As$ ridges during a self-aligned shadow mask step. The self-aligned metal nanowires at the Al_xGa_{1-x} As planes were subsequently transferred with the SNAP technique.

The spacer technique was used in an iterative way in [31] in order to define a nanomold yielding sub-10 nm nanowires with a 20-nm pitch. The process, *planar edge defined alternate layer (PEDAL)*, uses standard photolithography to define a sacrificial layer for the first spacer, then by iterating the spacer technique, a layer of dense spacers was defined. A shadow-mask deposited metal at the partially released spacer ridges formed thin nanowires, which were subsequently transferred onto the functional substrate. Since the spacer technique was used to define the nanomold and not the nanowires directly, this process is closer in nature to the nanomold-based techniques than to the spacer techniques.

2.1.2.5 Opportunities and Challenges of Top-Down Approaches

Top-down approaches are attractive because of their relatively easier required methods. The standard photolithography and spacer techniques can be integrated in a straightforward way into a CMOS process, whereas miscellaneous mask-based techniques may be more expensive and slower than the desired level for large production, and the maskless approaches may require the hybridization of the process with the non-conventional steps.

A promising opportunity that is offered by the spacer and nanomold-based techniques is the definition of devices with a sub-photolithographic pitch. In contrast to standard photolithography techniques, whose pitch is ultimately defined by the lithography limit, using spacer- and nanomold-based techniques represents an elegant way to circumvent the photolithographic limitations, and has a potential application field with regular architectures such as crossbar circuits as explained in Sect. 1.4.4.

The alignment of different processing steps is straightforward with standard lithography and miscellaneous mask-based techniques; while the spacer techniques are self-aligned. Nevertheless, whenever a nanomold-based step is introduced, the alignment becomes a very challenging issue, making these techniques more likely to be used at the early processing stages.

A general drawback of all these bottom-up techniques is that the obtained semiconducting nanowires are undifferentiated, meaning that the doping profile along the nanowires is generally the same and cannot be modified at later process stages after the nanowires are defined. In order to uniquely address every nanowire, it is highly desirable to associate a different doping profile to every nanowire in order to uniquely address them.

Another challenge that is specific to the nanomold-based technique is the metallic nature of the most demonstrated nanowires. However, in order to fabricate the access devices to the nanowires, it is required to have semiconducting nanowires that can be field-effect-controlled. Fortunately, there are still many opportunities promising the fabrication with semiconducting nanowires with these techniques as well.

2.2 Crossbar Technologies

In this part of the work, the goal is to consider technologies that enable the fabrication of nanowire crossbar circuits. Even though the nanowire fabrication techniques, surveyed previously, demonstrated their ability to yield layers of parallel nanowires, only a few of them were successfully used to demonstrate the feasibility of arrays of parallel nanowires in a crossbar fashion. In the following, demonstrated fabrication techniques for nanowire crossbars are surveyed, then potential crossbar switch technologies are explored.

2.2.1 Fluid-Directed Assembly

Nanowires fabricated with bottom-up processes have the property of generally being grown on a different substrate from the functional one. Consequently, they need to be dispersed into a solution and transferred onto the substrate to be functionalized. The iteration of the transfer operations with different directions may lead to a crossbar structure. In [19], layers of parallel nanowires were obtained by passing suspended nanowires in an ethanol solution through a fluidic channel structure formed between a PDMS mold and the flat substrate. The obtained results show a good alignment of the nanowires, which depends on the channel dimensions defined in the mold and the flow rate and duration. By patterning the substrate with NH_2-terminated regions, the alignment and density of nanowires could be improved. The second layer was obtained by using a crossed flow with respect to the direction of the first layer. The adhesion of the first nanowire layer to the substrate was demonstrated to be sufficiently strong that the sequential flow step did not affect the preceding one. The obtained crossbar structure shows a separation between the nanowires of about 400 nm.

2.2.2 Electric-Field-Assisted Assembly

Nanowires fabricated in a bottom-up process and suspended in a solution can be assembled in parallel layers by applying an electric field that directs their adhesion to the substrate while a flow of the suspended nanowires is applied to the substrate. In [32] this technique was applied on gold nanowires in order to align them parallel to the applied electric filed. The alignment was explained by the forces resulting from the polarization of the nanowire in the electric field. The approach was demonstrated in [33] to be feasible for doped semiconducting nanowires as well. The iteration of the same technique with two orthogonal directions of the electric field yields crossing nanowires in a crossbar fashion. However, the scalability of this method was evaluated in [19] to be limited by the electrostatic interference between nearby electrodes, and the requirement for an extensive lithography to fabricate the electrode for assembly of multiple nanowire arrays.

2.2.3 Nanomold-Based Nanowire Crossbars

Nanoimprint lithography was used in [34] in order to define two orthogonal layers of metallic nanowires. The nanomold was fabricated by electron-beam lithography and *reactive ion etching* (*RIE*) of a SiO_2-covered silicon substrate. The mold was then pressed onto a spin-coated *polymethylmethacrylate* (*PMMA*) on an oxidized Si substrate. After the system was heated, the mold was released resulting in a pattering of the PMMA, which was subsequently used as a lift-off mask to pattern a layer of parallel Ti/Pt nanowires. A layer of molecular switches, [2] rotaxane, was deposited over the entire substrate using the *Langmuir−Blodgett*(*LB*) method [35], in order to prepare the placement of the molecular switches at the crosspoints. Then, the fabrication of the top nanowire layer started with a thin Ti layer, which was very reactive with the top functional group of the molecules, and avoided the further penetration of the metal atom into the molecules. Then, a perpendicular top Ti/Pt nanowire layer was patterned in a similar way as explained for the low nanowire layer. The obtained functionalized crossbar had a crosspoint area of about 40×40 nm^2. The same approach was improved in [36] yielding a nanowire width of 30 nm and a pitch of 60 nm in a 1 kbit memory, which is equivalent to a crosspoint density of 28 Gbit/cm^2.

High-density crossbars were also demonstrated with the SNAP technique that was explained in Sect. 2.1.2 [37]. The fabricated nanowires were not only metallic but also silicon-based, with a width varying between 18 and 20 nm and a pitch between 30 and 60 nm. The iteration of the SNAP process was shown to be a flexible way to fabricate nanowire crossbars with densities up to 10^{11} cm^{-2} [30].

Such crossbars were functionalized with [2]rotaxanes and demonstrated for the first time 160-kbit molecular memories with a density of 10^{11} bit/cm^2 [38].

2.2.4 Crossbar Switches

While the scaling of the fabricated features below the photolithographic limit by applying the previously explained nanowire crossbar fabrication techniques has been successfully demonstrated, the interest in fabricating molecular devices, in which ultimately a small number of molecules is electrically connected to the electrodes, has continuously increased. The reason behind such an ambition, is not only the perpetuation of scaling power consumption and area per electronic device, but also the ability to understand the electrical behaviour of single molecules and to assess the physics of the zero-dimensional device in contact with a three-dimensional semiconducting or metallic bulk. In the meantime, the ability of modeling and synthesizing molecules that are potentially interesting for molecular electronics has shown many advances.

Many tries have been carried out in the last decades to design molecules comprising a donor-(σ bridge)-acceptor, which would have an asymmetric behaviour, allowing the current to flow in a preferential direction. Hereby, the electrode metal has to be carefully chosen, since the molecule-electrode interface was shown to interfere with the rectifying behavior in some cases [40–42]. Another class of switching molecules is represented by bistable molecules, such as [2] rotaxanes, pseudorotaxanes and [2] catenanes. They consist of two mechanically interlocked, or threaded, components. The molecule has two stable states (Fig. 2.6) and can be switched between these two states when the appropriate bias voltage is applied [39, 43].

Other research groups focused on phase change materials as a switching material at the nanowire crosspoints. In [44], a Ge nanowire pn-junction diode grown as a vertical nanowire with sublithographic dimensions was used as a memory cell. The cell is initially reset (to a low conductance state 0); then, it can be set buy applying a current as large as a few µA. At this point, the partially amorphous phase (in the reset state) is programmed into the poly-crystalline phase (SET state). In the SET state, the cell works as a diode. The programming from the SET to the RESET states is performed by applying pulses of large bias voltage. The cell provides an isolation of 100× between the forward and reverse bias in the SET state.

2.2.5 Comparison Between Crossbar Technologies

The main goal of the crossbar architecture is to organize nanowires into very dense arrays, with a pitch below the photolithographic pitch. The two crossbar approaches suggested for nanowires grown in a bottom-up process (i.e., fluid-directed and

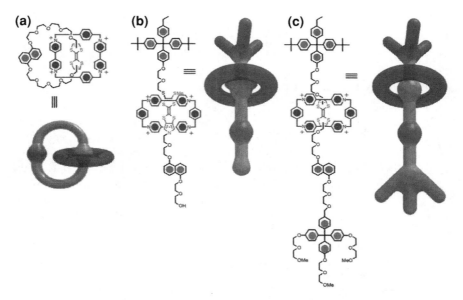

Fig. 2.6 Bistable mechanical molecular switches [39]: **a** [2] catenane, **b** pseudorotaxane and **c** [2] rotaxane

electric-field-assisted assembly) are not able to achieve this goal in general. The nanomold-based approach is the only existing technique that leverages the small dimensions of the nanowires and succeeds in arranging them into arrays with a sub-photolithographic pitch.

All these techniques showed promising results in terms of their ability to integrate nanowires into arrays, even though there is a certain discrepancy between the proposed approaches with respect to the nanowire dimensions and pitch. None of these techniques has targeted the exclusive utilization of CMOS processing steps in order to achieve crossbars with sub-photolithographic dimensions and pitches.

In the following sections, the actual work carried out with the fabrication facilities at EPFL is presented. The innovative goal of this work is to use a CMOS-compatible and cost-efficient technique, in order to arrange nanowires into arrays with a photolithography-*in* dependent pitch, while using only standard photolithography steps. The choices made with respect to the used equipments are justified in the following, before the fabrication steps and their optimization are explained.

2.3 Fabrication Facilities at EPFL

The CMI facilities were originally dedicated to the fabrication of *microelectromechanical systems* (*MEMS*) and to the optimization of their processing techniques. The available lithographic resolution of 0.8 μm is enough for this purpose. MEMS processing is technologically different from microelecronics fabrication techniques in the sense that MEMS processing may require the deposition of some metallic films in early stages of the process flow, which is forbidden in microelectronics processing, since such wafers would contaminate for instance high-temperature furnaces.

However, the CMI facilities are kept in a class-100 environment. All possible sources of contamination are avoided, offering a possible environment for microelectronics applications. The high-temperature furnaces are for instance kept free of any metallic or organic contamination, by systematically separating those dedicated to poly-Si deposition or oxide growth from those dedicated to the annealing of wafers already contaminated and by applying the usual *RCA* wafer cleaning procedure.

The CMI facilities were chosen to fabricate nanowire crossbars for different reasons. On the one hand, the microelectronics-compatibility is guaranteed in the CMI. On the other hand, the lithographic limit does not represent any stopper for the suggested fabrication process, since the nanowire width and pitch are inherently independent on the lithographic limit, and they only depend on the deposition thickness and the etching properties. Moreover, the CMI offers a flexible use of the equipments, including the mask writer, and a possible customization of the process recipes upon requests and following discussions between the CMI users and staff.

2.3.1 Photolithography

2.3.1.1 Mask Writing

The input file required by the lithography system is a CIF or GDSII file that maps the design layout drawn with any layout editor, such as *L-Edit* by *Tanner EDA* or *Virtuoso* by *Cadence*. Other less conventional input formats are acceptable as well. The ladder editor was used in this work with GDSII output files. Even though *L-Edit* may offer a user-friendly interface and some easier-to-use functionalities, *Virtuoso* is already well established within standard design flows in either large industrial projects or smaller academic designs.

The masks were written using the laser lithography system *Heidelberg DWL200*. The tool uses a laser scanner and a Krypton (Kr) laser source for g- and h-line photoresist, and it is suitable for batch processing. A lithography resolution of 2 µm was sufficient in this work, because of the independency of the process critical dimensions on the lithographic resolution. Even though a direct writing on the wafers is possible with this equipment, only glass masks (5" × 5" Cr-blanks) were written in this work, because this was more economical for a large number of processed wafers. Then, the masks were developed using *Süss DV* 10. The chromium and its native oxide were subsequently wet etched with the locally prepared solution $HClO_4 + Ce(NH_4)_2(NO_3)_6 + H_2O$ and the photoresist was stripped using the *Remover* by 1165 *Shipley Microposit*.

2.3.1.2 Photoresist

For non-lift-off steps, the *photoresist* (*PR*) that was used in this work is *AZ92xx*, which is a diluted version of *AZ9260* by *MicroChemicals*. The thickness of 2 μm was suitable to all photolithographic steps in this work, other than lift-off. *AZ92xx* offers the advantage of a high profile with vertical steps. The silicon substrate was spin-coated with the photoresist using *Rite Track* 88 *Series*, where the wafers were previously primed in a *YES* III *HMDS* primer oven at 150°C. During this priming process, *hexamethyldisilazane* (*HMDS*), $C_6H_{19}Si_2N$, was deposited on the wafer in order to enhance the adhesion of the photoresist to the silicon dioxide native film. The development of *AZ92xx* after exposure was carried outing using the same *Rite Track* 88 *Series*.

Lift-off steps were performed with a stack of two positive photoresists: *AZ1512* over *LOR*, both of them being supplied by *MicroChemicals*. For wafers with low profile topography (less than 0.7 μm), the thickness of *LOR* was set to 400 nm. The stack of both *AZ1512* and *LOR* was spin-coated in two successive steps and developed after exposure with *EVG* 150. For wafers with a high profile topography (higher than 1.6 μm), a single thick layer of *AZ92xx* (4–5 μm) was used instead of *AZ1512/LOR*, which was spin-coated and developed with *Rite Track* 88 *Series*. Following the development of the resist, the lift-off procedure and the photoresist strip were carried out at the *Plade Solvent Photolithography* wet bench (Z1).

2.3.1.3 Mask Alignment and Exposure

The mask alignment to the wafer and the exposure could be performed on either *Süss MA*150 or *Süss MA6/BA6*. In general, *Süss MA*150 was used for first masks in a batch mode, and *Süss MA6/BA6* was used for mask alignment with individual wafers.

2.3.2 Etching

2.3.2.1 Anisotropic Plasma Etch

The caves of both layers were defined inside silicon dioxide. The etching procedure requires the definition of vertical steps, which was performed with the plasma etcher *Alcatel AMS* 200 *DSE* using a fluorine chemistry (SF_6) with passivation gases for a better anisotropy. The definition of the multi-spacer had to fulfill the same requirements with respect to step anisotropy. Consequently, the oxide spacers were etched using the same equipment *Alcatel AMS* 200 *DSE*, while the poly-Si spacers were etched using a chlorine chemistry (Cl_2) with *STS Multiplex ICP* plasma etcher.

2.3.2.2 Isotropic Plasma Etch

The definition of the gate on a nanowire lying at the sacrificial layer needs to address the issue of a 3-dimensional (3D) gate following a step of about 0.5 μm height. The combination of anisotropic poly-Si etch to initially define the gate and a subsequent isotropic poly-Si etch in order to remove any unwanted residual poly-Si spacer at the sacrificial layer was necessary in this case, and was carried out using *Alcatel* 601*E* plasma etcher.

On the other hand, residual traces of photoresist needed to be completely stripped after the photoresist development, when the subsequent etch procedure is wet. This is known as a *descum* process, and the need for it is explained by the softness of the wet etching compared to the plasma etching, which may not be aggressive enough the remove the residual photoresist after development. An oxygen plasma etcher, *Oxford PRS*900, was utilized in order to isotropically etch the photoresist.

2.3.2.3 Wet Etch

In order to release the crossbar, the silicon dioxide between the two nanowire layers and in the sacrificial layer was wet-etched using a 7:1 *buffered HF (BHF)* solution in either *Coillard Etching* (Z6) or *Plade Oxide* (Z2) wet bench.

2.3.2.4 Chemical Mechanical Planarization (CMP)

In order to remove all residual films on the wafer back side, *Steag Mecapol E* 460 was used to polish the back side in a fast and straightforward way. Thereby, the front side was covered by a thick photoresist for protection, because the slurry used in CMP has nanometer scale SiO_2 particles that may stick at the wafer surface. It is therefore recommended to clean the wafer in a BHF bath after CMP operations.

2.3.3 *Thin Films*

The *Centrotherm* furnaces were used for poly-Si LPCVD (tube 1-1), Si_3N_4 LPCVD (tube 1-2), wet oxide growth (tube 2-2), gate oxide growth (tube 2-3), LTO LPCVD (tube 3-1), LTO and poly-Si densification (tube 2-1), Si doping with $POCl_3$ (tube 1-4), dopant diffusion (tube 2-4), and Cr/Ni annealing (tube 3-4).

The evaporation of the contact metal (Cr and $Ni_{0.8}Cr_{0.2}$) was performed with the *Alcatel EVA*600 evaporator. The equipment has *electron-beam* (*e-beam*) and thermal sources. For the present process flow, only e-beam sources were used. During evaporation, the vacuum level can be inside the evaporation chamber as low as 5×10^{-7} mbar. Such a level is lower than the obtained level during metal sputtering, making the evaporation a more attractive solution. The distance between the source and the wafer is around 0.5 m, which is lower than in other evaporation equipments. The disadvantage of a low distance is the possible anisotropy of the deposited metal film, that may deteriorated the quality of the lift-off process. Since the metal film was thin (around 50–60 nm), the lift-off process could be perfectly performed using *Alcatel EVA*600.

2.3.4 Wafer Cleaning

Following the different photolithography steps, the resist was stripped using the *Remover* 1165 by *Shipley Microposit*. Some plasma etching steps, namely the SiO_2 etch at 0°C with *Alcatel AMS* 200 *DSE*, harden the photoresist, which cannot be attacked by the *Remover* 1165 anymore. In this case, two successive steps of plasma oxygen etch (*Oxford PRS*900) alternated by a wet etch with the *Remover*1165 are performed.

The use of furnaces for poly-Si, Si_3N_4 and *low temperature oxide* (*LTO*) LPCVD necessitates a decontamination of the wafers from all organic or metallic residues. The usual and well established procedure is the RCA cleaning. It is performed at the *Plade RCA* (Z3) and consists in 3 baths. The first bath (RCA1) is $H_2O : NH_4OH : H_2O_2$ 5:1:1, used to remove organic residues; the second bath is $H_2O : HF$10 : 1, used to remove the native SiO_2; and the third bath (RCA2) is $H_2O : HCl : H_2O_2$ 6:1:1, used to remove the residual metal. The cleaning procedure can be transformed if the wafers have sensitive SiO_2 films or spacers, by replacing RCA1 by Piranha cleaning $H_2SO_4 + H_2O_2$ performed at the the *UltraFab* wet bench, and skipping the HF cleaning.

2.3.5 Process Control

Besides the continuous control of the wafer with optical microscopy (*Nikon Optiphot 200/Nikon Optiphot* 150) as well as *scanning electron-microscopy* (*SEM*) with *Zeiss LEO*1550, a regular control of the film thickness was conducted during the front-end phases, and particularly during the definition of the multi-spacer in order to optimize the deposition and the etching times. The measurements were done with the spectro-reflectometer *Nanospec AFT-6100*. Its resolution of about 10 nm was a limiting factor in certain cases. However, it was more convenient to use the *Nanospec AFT-*6100 to control film thicknesses, rather than cleaving the wafer and observing its cross-section with the SEM. An optical profiler was also utilized occasionally (*Veeko Wyko NT*1100).

Once the multi-spacer was defined, cross-sections were performed using the *FEI Nova* 600 *NanoLab*. The equipment has a dual beam: a SEM and a *focused ion beam (FIB)*. The FIB was used to cleave specific structures without damaging the rest of the wafer, while the SEM was used in imaging.

The electrical measurements represented and additional way to control the processed wafers in the back-end phases. A *Süss PM*8 manual prober station was available inside the CMI cleanroom for device characterization. However, these measurements were performed under (artificial) ambient light.

2.4 Process Flow

The goal of this part of the work is to define a process flow for a nanowire crossbar framework using standard CMOS processing steps and the available micrometer scale lithography resolution. The proposed approach is based on the spacer patterning technique presented in Sect. 2.1.2. The iteration of the spacer steps has been shown to be an attractive and cost-efficient way to fabricate arrays of parallel stripes used for the definition of nanomolds [31] or directly as nanowire arrays [28].

The approach presented in this part of the work is based on the idea of MSPT demonstrated in [28] for a single nanowire layer. In this part of the work, the efforts are concentrated on related challenges: first, the demonstration of the ability of this technology to yield a crossbar structure; then the assessment of the limits of this technology in terms of nanowire dimensions and pitch; and finally, the characterization of access devices operating as single poly-Si nanowire field effect transistors (poly-SiNWFET).

The main idea of the process is the iterative definition of thin spacers with alternating semiconducting and insulating materials, which result in semiconducting and insulating nanowires. The structures are defined inside a 1 μm high wet SiO_2 layer over the Si substrate (Fig. 2.7a). This SiO_2 layer has two functions: on the one hand, it insures the isolation between the devices; on the other hand it is used to define a 0.5 μm high sacrificial layer on which the multi-spacer is defined.

Then, a thin conformal layer of poly-Si with a thickness ranging from 40 to 90 nm is deposited by LPCVD in the *Centrotherm* tube 1-1 (Fig. 2.7b). During the LPCVD process, silane (SiH_4) flows into the chamber and silicon is deposited onto the substrate. The type of deposited silicon (amorphous or poly-crystalline) depends on the chamber temperature and pressure [45–47]. The deposition has been specifically optimized for the CMI facilities [48]. At the deposition temperature of 600°C, the LPCVD process yields poly-crystalline silicon. Thereafter, this layer is etched with the RIE etchant *STS Multiplex ICP* using a Cl_2 plasma, in order to remove the horizontal layer while keeping the sidewall as a spacer (Fig. 2.7c). As the densification of deposited silicon improves the crystalline structure [49], the poly-Si spacer is densified at 700°C for 1 hour under N_2 flow in the *Centrotherm* tube 2-1.

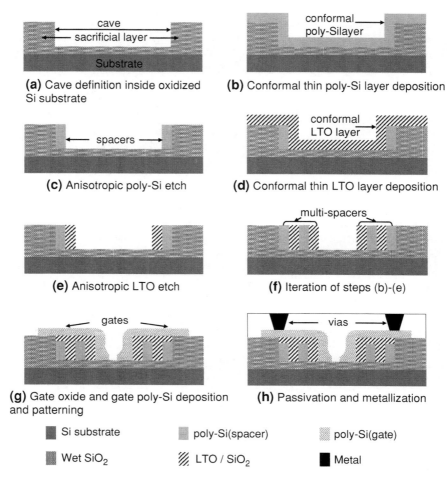

(a) Cave definition inside oxidized Si substrate

(b) Conformal thin poly-Si layer deposition

(c) Anisotropic poly-Si etch

(d) Conformal thin LTO layer deposition

(e) Anisotropic LTO etch

(f) Iteration of steps (b)-(e)

(g) Gate oxide and gate poly-Si deposition and patterning

(h) Passivation and metallization

| Si substrate | poly-Si(spacer) | poly-Si(gate) |
| Wet SiO$_2$ | LTO / SiO$_2$ | Metal |

Fig. 2.7 MSPT process steps

Then, a conformal insulating layer is deposited as a 40–80 nm thin LTO layer obtained by LPCVD in the *Centrotherm* tube 3-1 following the reaction of SiH$_4$ with O$_2$ at 425°C (Fig. 2.7d). The quality of the LTO can be improved through densification [50]. Thus, the deposited LTO is densified at 700°C for 45 minutes under N$_2$ flow. Then it is etched in the RIE etchant *Alcatel AMS 200 DSE* using C$_4$F$_8$ plasma in order to remove the horizontal layer and just keep the vertical spacer (Fig. 2.7e). Alternatively, instead of depositing and etching the LTO, the previously defined poly-Si spacer can be partially oxidized in the *Centrotherm* tube 2-1 in order to directly form the following insulating spacer.

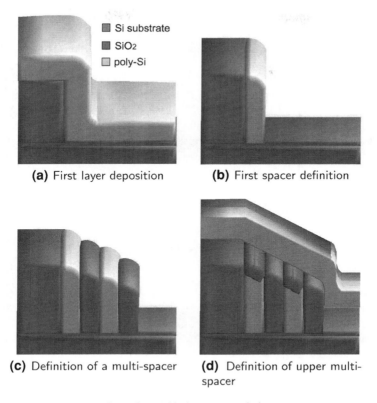

(a) First layer deposition   **(b)** First spacer definition

(c) Definition of a multi-spacer   **(d)** Definition of upper multi-
                                          spacer

Fig. 2.8 Main process steps of crossbars with the spacer technique

These two operations (poly-Si and insulating spacer definition) are performed one to six times in order to obtain a multi-spacer with alternating poly-Si and SiO$_2$ nanowires (Fig. 2.7f). Then, the batch is split into two parts: some of the wafers are dedicated to the definition of a second perpendicular layer of nanowires, some others are processed further with the gate stack and the back-end steps and are dedicated to perform electrical measurements.

In order to address the issue of characterizing a single access device (poly-SiNWFET), a single nanowire layer is used, on top of which a poly-Si gate stack is defined with an oxide thickness of 20 nm, obtained by dry oxidation of the poly-SiNW, and different gate lengths (Fig. 2.7g). The drain and source regions of the undoped poly-SiNW are defined by the e-beam evaporation and lift-off of 10 nm Cr and 50 nm nichrome (Ni$_{0.8}$Cr$_{0.2}$) with *Alcatel EVA* 600 (Fig. 2.7h). The Cr enhanced the adhesion and resistance of Ni to oxidation during the two-step annealing (including 5 minutes at 200°C, then 5 minutes at 400°C [51]) performed in the *Centrotherm* tube 3-4. Using Cr/Ni$_{0.8}$Cr$_{0.2}$ is a simple way to contact undoped nanowires, since Ni is a mid-gap metal. If the nanowires are doped, then it is possible to use aluminum as contact metal, which is then evaporated and patterned after a passivation layer is deposited and vias are opened, as depicted in Fig. 2.7h.

In order to address the issue of realizing a crossbar framework, the bottom multi-spacer is fabricated as explained previously in Fig. 2.7a–f, then a 20-nm dry oxide layer is grown as an insulator between the top and bottom nanowire layers. The top sacrificial layer is defined with LTO perpendicular to the direction of the bottom sacrificial layer. Then a poly-Si spacer is defined at the edge of the top sacrificial layer in a similar way to the bottom poly-Si spacers. Therafter, the separation dry oxide and both sacrificial layers are removed in a BHF solution in order to visualize the crossing poly-Si spacers realizing a small poly-Si nanowire crossbar. The main process steps of a crossbar are depicted in Fig. 2.8. In this figure, the difference in height between successive spacers is shown; which effect will be explained in the following sections.

A typical runcard of the process for a single layer of nanowires is detailed in Table 2.1. When a double layer is targeted by the process (i.e., a crossbar), then additional photolithography and spacer definition steps are included between steps 4 and 5 in Table 2.1. These extra steps correspond to the upper sacrificial layer. In this typical runcard, the thickness of poly-Si and LTO layers are just given as examples. These values can be varied within the processed wafers. This runcard shows the steps for Cr/$Ni_{0.8}Cr_{0.2}$ contacts defined with lift-off. Some wafers were contacted with Al through vias. The annealing can be done either after every spacer definition as explained above, or it can be done once all spacers are defined, as shown in Table 2.1.

2.5 Process Optimization

The development of a new technology necessitates the optimization of a large number of parameters, even though the abilities of the equipments and the goal set to use CMOS compatible steps put some boundary conditions that reduce the range of fabrication parameters to be explored. In this section, some process optimization aspects are highlighted and the adopted solutions are explained.

2.5.1 Etch of Sacrificial Layers

The process has two perpendicular sacrificial layers, corresponding to each one of the crossing nanowire layers. The first sacrificial layer is etched in a wet oxide at 0°C with a SF_6 plasma, using *Alcatel AMS 200 DSE*. It is required that the etch is highly anisotropic in order to obtain a vertical step. The shape of the step defines the shape of the conformal poly-Si thin film. A vertical step would result in a removal of only lateral parts of the poly-Si film, leaving just the narrow vertical spacer. In reality, the step has a certain obtuse angle with the horizontal line, thus

Table 2.1 Typical run card for a single nanowire array

Step	Process	Equipment	Recipe
1	*Wafer oxidation*		
1.1	RCA cleaning	Z3/RCA wetbench	Standard RCA1+HF+RCA2
1.2	Wet oxidation	Z3/Centrotherm 2-2	1 μm
1.3	Inspection	Z3/Nanospec AFT-6100	
2	*Photolithography: sacrificial layer mask*		
2.1	Wafer priming	Z1/YES III, HMDS primer oven	$\sim$ 20 min
2.2	PR coating	Z1/Rite Track 88 series	2 μm of AZ92xx
2.3	PR exposure	Z1/MA/BA6	9 s, hard contact
2.4	PR development	Z1/Rite Track 88 series	2 μm of AZ92xx
3	*Definition of sacrificial layer*		
3.1	RIE oxide etch	Z2/AMS 200 DSE	SiO$_2$_PR_5:1 2'30"
3.2	O$_2$ PR strip	Z2/Oxford PRS900	1 h
3.3	PR wet strip	Z2/WB PR strip .	Standard 2 × 5 min
3.4	O$_2$ PR strip	Z2/Oxford PRS900	10 min
3.5	Inspection	Z3/Nanospec AFT-6100	
4	*Definition of the multi-spacer*		
4.1	Piranha cleaning	Z2/WB Piranha	Standard 2 × 5 min
4.2	RCA2 cleaning	Z3/WB RCA	Standard 15 min
4.3	Poly-Si LPCVD	Z3/Centrotherm 1-1	80 nm
4.4	Inspection	Z3/Nanospec AFT-6100	
4.5	Poly-Si etch	Z2/STS Multiplex ICP	Sub_Si 0'19"
4.6	Inspection	Z3/Nanospec AFT-6100	
4.7	Piranha cleaning	Z2/WB Piranha	Standard 2 × 5 min
4.8	RCA2 cleaning	Z3/WB RCA	
4.9	LTO LPCVD	Z3/Centrotherm 3-1	90 nm
4.10	Inspection	Z3/Nanospec AFT-6100	
4.11	Oxide etch	Z2/AMS 200 DSE	SiO$_2$_PR_5:1 0'27"
4.12	Inspection	Z3/Nanospec AFT-6100	
	Iteration of steps 4.1-4.12 n ×		
4.($12n + 1$)	Densification	Z3/Centrotherm 2-1	45' at 700°C under N$_2$ flow
4.($12n + 2$)	Cross-section	Z8/FEI Nova 600 NanoLab	FIB cross-section and SEM inspection
5	*Photolithography: cave edge mask*		
5.1	Wafer priming	Z1/YES III, HMDS primer oven	$\sim$ 20 min

(continued)

Table 2.1 (continued)

Step	Process	Equipment	Recipe
5.2	PR coating	Z1/Rite Track 88 series	2 µm of AZ92xx
5.3	PR exposure	Z1/MA/BA6	9 s, hard contact
5.4	PR development	Z1/Rite Track 88 series	2 µm of AZ92xx
6	*Cave edge etch*		
6.1	Descum	Z2/Oxford PRS900	Plasma O_2 for 0'30"
6.2	Wet oxide etch	Z2/WB oxide etch	BHF 30"
6.3	Native oxide strip	Z2/Alcatel 601E	SiO_2_Soft 0'10"
6.4	Poly-Si strip	Z2/Alcatel 601E	Si_Iso_Slow 2'00"
6.5	PR wet strip	Z2/WB PR strip	Standard 2 × 5 min
6.6	O_2 PR strip	Z2/Oxford PRS900	30 min
6.7	Inspection	Z3/Nanospec AFT-6100	
7	*Gate stack deposition*		
7.1	Native oxide strip	Z6/WB Oxide etch	BHF 30"
7.2	Piranha cleaning	Z2/WB Piranha	Standard 2 × 5 min
7.3	RCA2 cleaning	Z3/WB RCA	Standard 15 min
7.4	Dry oxidation	Z3/Centrotherm 2-3	20 nm gate oxide
7.5	Poly LPCVD	Z3/Centrotherm 1-1	0.5 µm deposition of gate poly-Si
8	*Photolithography: gate mask*		
8.1	Wafer priming	Z1/YES III, HMDS primer oven	∼20 min
8.2	PR coating	Z1/Rite Track 88 series	2 µm of AZ92xx
8.3	PR exposure	Z1/MA/BA6	9 s, hard contact
8.4	PR development	Z1/Rite Track 88 series	2 µm of AZ92xx
9	*Gate patterning*		
9.1	Anisotropic poly-Si etch	Z2/STS Multiplex ICP	Sub_Si 1'20"
9.2	Isotropic poly-Si etch	Z2/Alcatel 601E	Si_Iso_Slow 0'20"
9.3	O_2 PR strip	Z2/Oxford PRS900	1 h
9.4	PR wet strip	Z2/WB PR strip	Standard 2 × 5 min
9.5	O_2 PR strip	Z2/Oxford PRS900	10 min
10	*Photolithography: metallization mask*		
10.1	Dehydration	Z6/Memmeret dry box	20 min at 150°C
10.2	PR coating	Z6/EVG150	400 nm of AZ1512 on LOR (standard recipe 5_5)
10.3	PR exposure	Z6/MA/BA6	1.5 s
10.4	PR development	Z6/EVG150	400 nm of AZ1512 on LOR (standard recipe 5_5)
11	*Metallization*		
11.1	Native oxide strip	Z6/WB Oxide etch	0'30"

(continued)

Table 2.1 (continued)

Step	Process	Equipment	Recipe
11.2	Metal evaporation	Z4/Alcatel EVA 600	10 nm Cr + 50 nm $Ni_{0.8}Cr_{0.2}$
12	*Lift-off*		
12.1	Lift-off	Z1/WB Photolitho	7 h
12.2	Sonication	Z1/WB Photolitho	5 min
12.3	PR strip	Z6/WB PR strip	Standard 2 × 5 min
13	*Thermal annealing*		
13.1	Annealing	Z3/Centrotherm 3-4	5' at 200°C + 5' at 400°C in N_2 flow
14	*Back-side CMP*		
14.1	Wafer priming	Z1/YES III, HMDS primer oven	~20 min
14.2	Font-side PR coating	Z1/Rite Track 88 series	2 µm of AZ92xx
14.3	Back-side CMP	Z5/Steag Mecapol E 460	3 × 3' + SRD, slurry = Reclaim
14.4	PR strip	Z6/WB Resist strip	Standard 2 × 5 min

the conformal poly-film is partially removed in the vertical direction as well. In some cases, this can lead to a serious deformation of the poly-Si spacer shape, which is carried forwards to the next spacers as well.

The SF_6 etch process can be improved by adding a passivation gas (CH_4), which enhances the verticality of the step by passivating the areas as soon as they are etched. A comparison of the results with and without passivation gazes is demonstrated in Figs. 2.9 and 2.10 respectively with SEM of two structures processed differently.

2.5.2 Spacer Definition

The spacer etch procedures have to be anisotropic in order to keep the spacer shape as vertical as possible as explained above. This was solved for the oxide etch by using a SF_6 plasma etch including CH_4 passivation, in a similar way to the

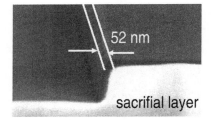

Fig. 2.9 SEM of poly-Si spacer defined on anisotropically etched SiO_2 sacrificial layer (passivated SF_6 etch)

Fig. 2.10 Cross-section of poly-Si spacer defined on slightly isotropically etched SiO$_2$ sacrificial layer (bare SF$_6$ etch). Spacer thickness ~60 nm

Fig. 2.11 Oxide spacer over-etch: first oxide spacer stripped, leaving 2 neighboring poly-Si spacers. Second oxide spacer over-etched, leading to etch of the cave (*circle*)

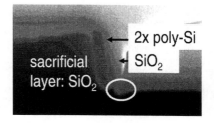

sacrificial layer etch. For the poly-Si etch, the used chlorine chemistry (with just Cl$_2$ plasma) was sufficient to insure the desired anisotropy.

An additional constraint on the spacer etch is the accuracy of etch time. As a matter of fact, an over-etch of any conformal layer (either in poly-Si or in LTO) reduces the thickness of its vertical part after the lateral part is fully consumed. In addition, the over-etch of a conformal layer attacks the uncovered layer underneath it, which is the SiO$_2$ sacrificial layer (Fig. 2.11). This is clearly more critical when the etched layer is LTO. Even the poly-Si layer over-etch may be of concern though, since the selectivity of the Si etchant to the SiO$_2$ etchant is worse than 1:20, which may result in up to 5 nm of loss of the sacrificial layer. The loss is cumulative over the whole multi-spacer definition, and becomes more visible with a larger number of spacers, resulting in an oblique multi-spacer, as demonstrated in Fig. 2.12. Even though such a shape may be processed further, it is not desirable to have an oblique gate on top of the nanowires because of possible induced stress.

On the other hand, an under-etch of the conformal layer results in the opposite effect. If a LTO is under etched, then more materials are deposited and left inside the cave of the sacrificial layer, resulting in turns in an oblique shape as well.

Fig. 2.12 Oxide and poly-Si spacers over-etch: all oxide spacers are stripped, poly-Si is over-etched. Then, cave is attacked resulting in diamond-shaped multi-spacer

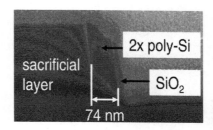

Fig. 2.13 Calibration of the poly-Si etch with Cl_2 chemistry. The offset is due to the etch of the native oxide

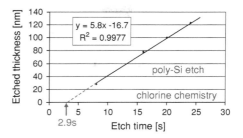

Fig. 2.14 Calibration of the SiO_2 etch with SF_6 chemistry. The etch rate is constant and there is almost no offset

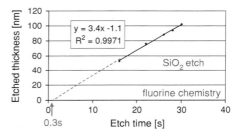

The under-etch of the poly-Si conformal layer leaves some poly-Si grains inside the cave and may result in a short between the different poly-Si spacers.

In order to ovoid both under- and over-etch of poly-Si and LTO, the etch rates are accurately calibrated with both fluorine and chlorine chemistries for LTO and poly-Si etch respectively. The etch procedures are calibrated by measuring the poly-Si and LTO layers before and after partial etching of thin poly-Si and densified LTO layers using the *Nanospec AFT*-6100 and different etching times. The etching times are chosen such that the remaining film is thicker than the *Nanospec AFT*-6100 measurement limit of 10 nm.

Figure 2.13 shows the calibration plot for the chlorine chemistry used to etch poly-Si. A linear curve with an offset of about 2.9 s is clearly seen in this plot. The offset is due to the fact that the Cl_2 plasma first physically attacks the native SiO_2, before the poly-Si is attacked. On the contrary, Fig. 2.14, shows that the calibration plot for the fluorine chemistry used to etch LTO densified at 700°C for 45 min under N_2 flow is almost offset-free and linear with an etch rate of about 206 nm/min.

2.5.3 Gate Stack

Some samples were used in order to electrically characterize the access devices operating as poly-SiNW. In this case, a gate stack was defined on top of the poly-Si spacer following the step defined by the sacrificial layer with a height of

Table 2.2 Etch time for 3 calibration recipes. Anisotropic and isotropic etch are carried out with *STS Multiplex ICP*/Sub_Si and *Alcatel* 601*E*/Si_Iso_Slow respectively

Recipe	1	2	3
Anistropic etch	1'20"	1'20"	1'20"
Isotropic etch	0'10"	0'30"	0'20"

0.4–0.5 μm. An anisotropic etch of the gate poly-Si would leave a poly-Si spacer at the sacrificial layer shorting the drain and source of the device. Thus, an additional isotropic gate poly-Si etch procedure was required in order to remove this parasitic spacer. The isotropic etch should just remove ∼0.4μm laterally from the gate edge. An over-etch may strip either the gate or the underlying poly-SiNW. An under-etch causes a shirt between drain and source. This combination of two types of etching being crucial for the device operation, several etch times were tried as summarized in Table 2.2. The impact of the double etch is illustrated with the SEM images in Fig. 2.15.

2.6 Device Characterization

The process is optimized according to the steps described above in order to carry out different types of investigation. First, a structural characterization was conducted in order to assess how small the nanowire pitch and how dense the nanowire crosspoints can be achieved. Then, an electrical characterization of single nanowires with drain, source and gate contacts was performed in order to assess the ability of a single poly-SiNWFET to act as an access transistor to the underlying nanowire.

2.6.1 Structural Characterization

In the following, the structural properties of arrays of parallel nanowires fabricated with the proposed technique are assessed. Figure 2.16 shows a SEM image of 3 double-spacers poly-Si/LTO. All the poly-SiNW have a uniform thickness of 54 nm, indicating that the NW thickness can be accurately controlled. Their height was about the height of the sacrificial layer, and decreased with increasing number of spacers, because of the increasing number of etching procedures. A NW length of hundreds of micrometers could be achieved, with no NW breakage. Besides the advantage of exclusively using standard CMOS steps, this technique has a high yield close to unity.

The insulating nanowires can be made either with LTO deposition or dry oxidation of the previously defined poly-Si nanowires. The second option was investigated by oxidizing about 40 nm of every poly-SiNW in a nanowire layer

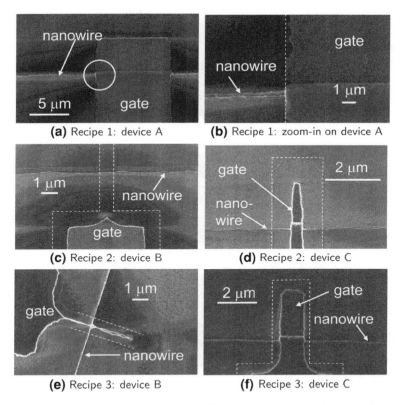

(a) Recipe 1: device A

(b) Recipe 1: zoom-in on device A

(c) Recipe 2: device B

(d) Recipe 2: device C

(e) Recipe 3: device B

(f) Recipe 3: device C

Fig. 2.15 Optimization of 3D gate etch. Dashed lines are the gate dimensions as defined by the photoresist (showed in order to guide the eyes). **a** Isotropic etch time is too short: gate edge just starts to be laterally etched. **b** Zoom-in on the circle in **a**: drawn gate edge is locally not etched. **c** Over-etch of the gate in shortest devices (1 μm). **d** Longer gates (2 μm) survive the lateral gate over-etch. **e** Shortest device (1 μm) are laterally just etched the required amount ∼0.4 μm. **f** Longer gates (2 μm) suggest that the combination of anisotropic and isotropic etch is satisfactory

Fig. 2.16 SEM of a poly-Si/LTO 6×-spacer: poly-Si spacers have uniform thickness of ∼54 nm

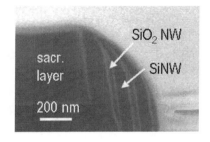

containing 6 poly-SiNW, yielding a theoretical dry oxide thickness about 90 nm.
A SEM image of the cross-section of the obtained structure is illustrated in
Fig. 2.17, showing an average width of the obtained poly-SiNW of about 60 nm.
The surface roughness of the obtained structures is coarse because of the non-
uniform oxidation rate of the poly-Si at the grain boundaries. Consequently, the
deposition of LTO is found to be a better way to insulate the nanowires than the
dry oxidation of poly-Si, even though it necessitates more processing steps.

The scalability of this technique was investigated by depositing thinner poly-Si
layers (40 nm): Fig. 2.18 shows that the obtained poly-SiNW have a width of
20 nm. For the device in this SEM image, the multi-spacer was planarized using
CMP after it was defined. This result demonstrated that it is possible to make the
nanowires narrow by depositing less poly-Si. The obtained nanowire width is
always less than the deposited film thickness, because of the etch procedure that
attacks the vertical structures in a measurable way, but much less than it attacks
the lateral structures.

The possible use of the MSPT for the fabrication of two perpendicular layers of
crossing NW is illustrated in Fig. 2.19 with one poly-SiNW crossing 4 poly-SiNW
underneath it. Here gain, the length of the nanowires in the crossbar could be made
as large as desired without any noticeable nanowire breakage. The obtained
crosspoint area is about $100 \times 100\,\mathrm{nm}^2$, which is equivalent to a crosspoint density
of $10^{10}\,\mathrm{cm}^{-2}$.

Fig. 2.17 SEM of poly-Si/
dry SiO$_2$ 12×-spacer: the
spacers are repeatable with a
large number of iterations

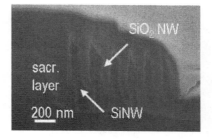

Fig. 2.18 SEM of an ultra
thin multi-spacer: poly-SiNW
thickness ∼40 nm

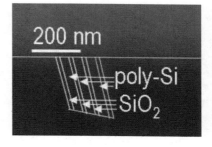

Fig. 2.19 SEM of a small
4 × 1-crossbar. The upper
layer is separated by 20 nm
from the lower layer

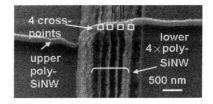

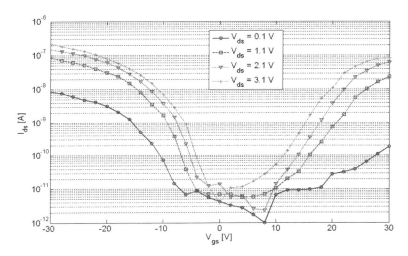

Fig. 2.20 $V_{gs} - I_{ds}$ curve of undoped poly-SiNWFET with a back gate and nickel silicide
contacts

2.6.2 Electrical Characterization

The need to access the nanowires and control the current flow through them
motivates for the definition of access transistors having a poly-Si spacer as a
channel. Undoped poly-SiNWFET with a single poly-Si spacer as a channel are
characterized. The $I_{ds} - V_{gs}$ curves for back-gated devices show an ambipolar
behaviour, with a current conductance under either high positive or negative gate
voltage (Fig. 2.20). The ambipolar behaviour is explained by two factors:
the undoped poly-SiNWFET channel and the mid-gap drain and source nickel
silicide metal, resulting in a Schottky barrier at the drain and source contacts
whose thickness can be modulated using the electrical gate field (Fig. 2.21),
making it more transparent for either electrons or holes. A similar behaviour in
other SiNW technologies has already been reported and explained by the existence
of a Schottky barrier at the drain and source contacts [52–54].

The I_{on}/I_{off} ratio was about 2×10^4 and 4×10^3 for the p- and n-branch
respectively. The low $I_{on} = 0.2\,\mu\text{A}$ and $0.1\,\mu\text{A}$ for p- and n-branches respectively is
explained by the low W/L ratio (the nanowire width $W = 67$ nm, back gate length

Fig. 2.21 Band diagram of
ambipolar undoped poly-
SiNW devices. *Left* positive
V_{gs} makes Schottky barrier
transparent for electrons.
Right negative V_{gs} makes
Schottky barrier transparent
for holes

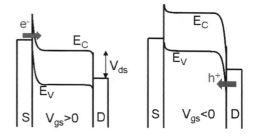

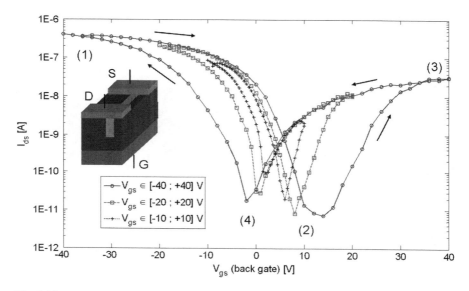

Fig. 2.22 Variation of hysteresis width of $I_{ds}-V_{gs}$ curve and I_{on}/I_{off} ratio of back-gated devices
with V_{gs} range

$L = 20 \, \mu m$, gate oxide thickness $t_{ox} = 400 \, nm$) and the low mobility in poly-Si.
The Schottky barrier for holes may be slightly lower than for electrons, which
explains the higher I_{on}-current in the p-branch. The curves were repeatable on the
wafer scale and they do not represent a single device. The ability to control the
devices in a FET fashion proves their possible use as access devices to the NW layer
within a decoder. The ambipolarity is due to the intrinsic channel and the mid-gap
contact metal that is electrostatically controlled by the gate field. By using implanted
contact regions and metal contact, the unipolar behavior can be restored [53].

The ambipolarity of a single poly-SiNW is investigated for different sweep
ranges of V_{gs}. The $I_{ds} - V_{gs}$ curve in Fig. 2.22 shows a hysteric behavior when V_{gs}
is swept from $-V_{gs,max}$ to $+V_{gs,max}$ back and forth. By enlarging the V_{gs} sweep
range from $V_{gs,max} = 10 \, V$ to $40 \, V$, the hysteresis became larger. Charge trapping

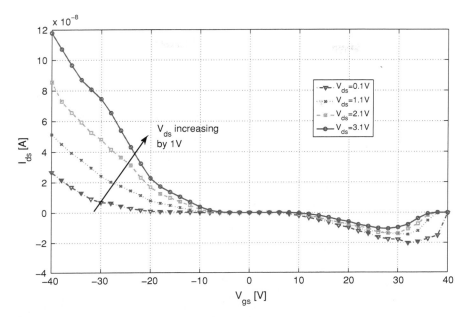

Fig. 2.23 $I_{ds}-V_{gs}$ curve of undoped poly-SiNWFET with top gate ($L_g = 4\,\mu m$, $W_{NW} = 67\,nm$ and $t_{ox} = 0.450\,\mu m$)

and detrapping can explain the hysteresis[1] in Fig. 2.22. From (1) to (2), trapped holes at the SiO_2/poly-Si interface create positive fixed charges, and are detrapped with increasing V_{gs}. From (2) to (3), an electron channel is created. With the increasing electron density, more electrons are trapped at the SiO_2/poly-Si interface, leading to a negative interface charge density. From (3) to (4), electrons are detrapped with the vanishing electron channel. Detrapping is a slower process than trapping, explaining the hysteresis (2)-(3)-(4). From (4) to (1), a hole channel is created and increases the trapping probability for holes. This is a faster process than detrapping holes; explaining again the hysteresis (4)-(1)-(2). The off-current at (2) is lower than the one at (4) because of the additional probability of having electrons trapped in SiO_2 besides the electrons trapped at the SiO_2/poly-Si interface. By enlarging the sweep range (higher V_{gs}), more charge carriers are trapped. This shifts the threshold voltage in the n-branch (p-branch) to more positive (negative) values during the trapping phase. The detrapping is slow, thus all 3 curves almost coincide during the detrapping phases.

Devices with a top-gate showed a unipolar behaviour with a p-type polarity despite the intrinsic channel (Fig. 2.23). The unipolar behaviour is explained by the absence of an electrostatic control of the gate on the drain/source-to-channel

---

[1] This conclusion is qualitatively confirmed with Silvaco device simulations performed by Dr. Nikolaos Archontas, Democritus University of Thrace, Greece

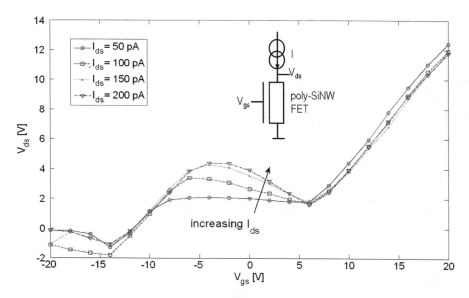

Fig. 2.24 V_{ds} versus V_{gs} curve of undoped poly-SiNWFET for fixed I_{ds} with a back gate and nickel silicide contacts

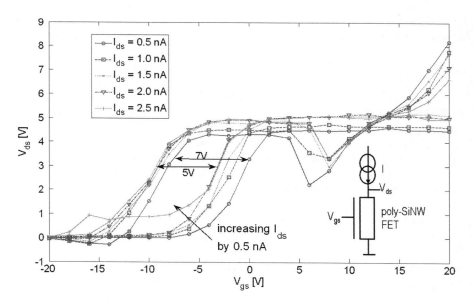

Fig. 2.25 Hysteresis of the $V_{ds} - V_{gs}$ transfer characteristics

contact area [52], which is responsible for the modulation of the Schottky barrier and the resulting hole and electron currents. The dominance of the p-type behaviour confirms the fact that nickel silicide contacts have a lower Schottky for holes than for electrons.

The transfer characteristics $V_{ds} - V_{gs}$ is shown in Fig. 2.24 for a fixed I_{ds}, and has a clear negative slope region. The same transfer characteristics have a hysteresis of 5–7 V, which decreases with increasing injected current I_{ds} (Fig. 2.25). The measured hysteresis is in agreement with the behaviour of poly-SiNW reported in literature and it can be explored in single nanowire memories [55].

2.7 Potential Applications

2.7.1 Crossbar Structures

A promising application of SiNW is the fabrication of crossbar structures. Previous approaches to build NW crossbars achieved *i*) metallic arrays, which do not have any semiconducting part that can be used as an access transistor, or *ii*) silicon-based crossbars with fluidic assembly, which have a larger pitch in average than the photolithography limit. Table 2.3 surveys the reported realized crossbars and shows that the proposed patterning technique has both advantages of yielding semi-conducting NW and a high crosspoint density $\sim 10^{10}$ cm^{-2}, as measured in the small crossbar of Fig. 2.19, while using conventional photolithographic processing steps. The use of the densest layers (Fig. 2.18) would yield a higher crosspoint density of 6.3×10^{10} cm^{-2}.

2.7.2 Single Poly-Si Nanowire Memory

Besides the application as a crossbar array, there is another potential application as poly-SiNW memory based on the hysteresis of the $V_{ds} - V_{gs}$ transfer characteristic for a fixed I_{ds}. A poly-SiNW memory cell was already proposed in [55] and its

Table 2.3 Survey of reported nanowire crossbars. Functionalized arrays are those including molecular switches

References	[38]	[56]	[57]	This work
NW material	Si/Ti	Ti/Pt	Si	poly-Si
NW width [nm]	16	30	20	54
NW pitch [nm]	33	60	>1000	100
Crossbar density [cm^{-2}]	10^{11}	2.7×10^{10}	N/A	10^{10}
Technique	SNAP	NIL	Self-assembly	MSPT
Functionalized?	yes	yes	no	no

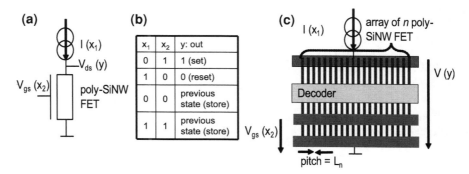

Fig. 2.26 Single poly-SiNW memory after [55]: **a** Poly-SiNW memory cell. **b** Memory operation. **c** High-density realization concept of poly-SiNW memory cells with MSPT

operation was experimentally demonstrated using I_{ds} and V_{gs} as inputs and V_{ds} as output storing the information. A single poly-SiNW memory cell after [55] and the organization of an array of similar cells in the proposed technology are illustrated in Fig. 2.26. The MSPT offers a denser pitch independent of the lithography, with a memory density scaling as L_1/L_n, where L_1 and L_n represent the lithography and the MSPT pitch respectively. At the 65 nm technology node ($L_1 = 2 \times 65$ nm) and with 20 nm wide nanowires ($L_n = 2 \times 20$ nm), the density with the MSPT is 3.2× higher. The density increases with decreasing Ln which scales with the thickness of the conformal layers.

2.7.3 Memristors

Memristors are the element that completes the symmetry of the equation system describing the relation between the fundamental circuit variables: current I, voltage V, charge Q and magnetic flux Φ [58]. They are the physical implementation of devices controlled by charge and magnetic flux and described by their memristance $M = d\Phi/dQ$. If the memristance is not constant, rather depending on Q, then the device $I - V$ curve shows a hysteric behavior that can be used in circuit design.

The significance of memristors arises from their natural existence in biological computational systems. For instance, ion diffusion is responsible of the time-dependant conductance of the neuron membrane in the Hodgkin-Huxley model [59], which is modeled as a memristive device. Learning mechanisms are also explained using the memristive model of synapses [60] and they were demonstrated with single devices [61]. Self-programming circuits were demonstrated by embedding memristors into logic circuits [62]. The memristor effect, seen as the hysteric behavior of the $I - V$ curve, may be obtained by using a circuitry that includes active elements and an internal power source [58, 63]. However, this solution is just an emulation of memristors and it consumes valuable chip area and power. With the scaling of device dimensions, new phenomena have been claimed to be responsible for the memristor behavior in monolithic devices. For instance, it has been demonstrated that a memristor effect arises naturally in nanoscale systems in which solid-state electronic and ionic transport are coupled under an external bias voltage [64]. Voltage driven memristor effects were demonstrated as well in Pt/organic/Ti [65], in *polyethylene oxide (PEO)/polyanilin (PANI)* polymeric [61] and in amorphous silicon [66] devices.

Unlike basic two-terminal memristors, the fabricated devices have three terminals. The channel conductance depends on the trapped charges; thus the channel resistance may be represented as a function of the charge, in a similar way to a two-terminal memristor. The gate offers the opportunity to control the resulting $I_{ds} - V_{ds}$ hysteresis in a field-effect fashion. The device can be exploited in order to save information, in a reminiscent way to what has been explained before for single poly-SiNW memory cells.

2.7.4 Nanowire Decoders

Fabricating crossbars with a sub-photolithographic pitch raises the question of how to make every nanowire addressed by the outer CMOS circuit through a decoder. The design of crossbar decoders has attracted a lot of attention and the proposed solutions are either analog [28, 67, 68] or digital [56, 57, 69, 70]. Crossbars fabricated in the proposed approach can be addressed in either ways. In [28], it was suggested that the doping level modulation in the poly-SiNW body can differentiate the nanowires and make the implementation of an analog decoder in the MSPT technology possible. However, this approach needs more signal processing during the READ operation. In the following, a novel concept of fabricating a fully deterministic digital MSPT decoder is presented. The design aspects related to this technique are addressed in Sect. 3.4.

While a pattern can be easily defined during the growth of nanowires in a bottom-up process (Sect. 2.1.1), it is more difficult to define it with top-down processes. For instance, the MSPT, yields a regular array of undifferentiated nanowires if the bare procedure depicted in Fig. 2.7 is applied. Once the array is defined on a sub-lithographic scale, it is difficult to pattern it with standard photolithographic means, unless expensive high-resolution and time-costly methods, such as electro-beam lithography, are applied. Consequently, it is desirable to pattern the nanowires while they are defined: i.e., whenever a new spacer is defined, it has to be patterned before the next spacer is defined.

The fabrication flow that includes the decoder is illustrated in Fig. 2.27 and should be understood as an extension inserted between steps depicted in Fig. 2.7b and f. Other steps remain unchanged. The additional steps are lithography patterning and doping after every spacer definition step. Specific regions from every poly-Si nanowire are defined and doped in this way. Nanowires are fragile and thin structures, and they should be doped carefully with light doses. However, the total

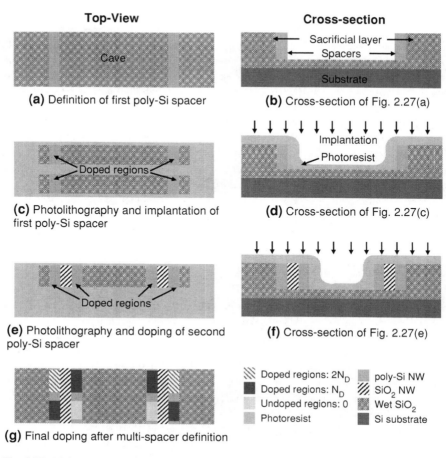

Fig. 2.27 Main process steps of MSPT decoder fabrication

doping level of a specific region is the sum of all doping levels accumulated in this region throughout the definition of the whole array, as illustrated in Fig. 2.27g. An optimized choice of the lithography/doping sequences and the doping doses may result in the desired nanowire pattern.

2.8 Discussions

One important question that may arise when it comes to the MSPT is the cost of the additional conformal deposition and RIE etch steps. The fabrication time needed for a 256×256 nanowire crossbar (8 kB memory) would be tremendous if 2×256 deposition/etch operations were required. Fortunately, the MSPT has two advantages. First, it can be parallelized within a single wafer: i.e., by using n parallel

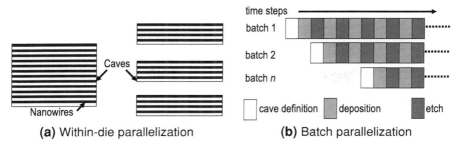

Fig. 2.28 Parallelization of the MSPT. **a** Using many small caves instead of a few large ones minimizes the number of steps, but has a cost in terms of area. **b** Any two batches can be processed together during the spacer definition steps, as long as the spacer parameters are identical

sacrificial layers instead of one, the number of deposition/etch steps is divided by n (Fig. 2.28a). Second, the technique allows for parallel batch processing, i.e., any two different batches can be processed together during the deposition/etch steps as long as the thickness of the conformal layers is the same (Fig. 2.28b).

Another important question about the proposed technique is related to the lower mobility of current carriers in the poly-Si used to define the structure, compared to crystalline Si. The question was generalized previously for any crossbar type: whatever the used NW material, the structure length and small cross-section will induce a slower signal propagation and higher resistance. To address this fact, it is generally believed [71] that the benefit of crossbars is to parallelize memory and computation in a grid with a large number of small crossbars, rather than using a limited number of large crossbars.

Considering the different potential applications presented in this chapter, it may be interesting to think of the advantages of using this process in order to fabricate single poly-SiNW memories or memristors, given the wide range of technologies that can be deployed to fabricate these devices. Clearly, the benefit in terms of area and integration density is the dominant advantage, since the considered technique easily yields structures with a pitch below the photo-lithographic limit. However, it is questionable to separately access these devices from the rest of the circuit defined on the lithography scale. This access requires the use of a decoder that addresses every nanowire separately. The benefits of this technology cannot be explored unless a decoder is available. The design aspects and the fabrication complexity of the proposed decoder concept will be therefore the core of the following chapter.

2.9 Chapter Contributions and Summary

This chapter has presented a first-time demonstration of the opportunity to use the multi-spacer patterning technique in order to fabricate the framework for crossbar circuits by integrating two layers of poly-Si spacers on top of each other.

The technique is cost-efficient and CMOS-compatible, in the sense that it does not require any additional steps other than those used in a standard CMOS process. It is also the only technique that uses only photolithography steps, while it is able to define structures with dimensions and pitch well below the photolithography pitch.

It has been demonstrated that the benefits of this technology go beyond the limit of crossbars circuits. In principle, the sub-photolithographic half pitch can be beneficial in any other application of the fabricated poly-SiNWs in terms of integration density. The hysteresis of the fabricated structures suggests their use as single poly-SiNW memories or memristors.

In this chapter a novel concept of fabricating nanowire decoders with the multi-spacer patterning technique has been introduced. It is the first time that a decoder for this nanowire technology is proposed. It has the advantage of being digital and deterministic. It is expected to have a minimal cost in terms of size. The technological costs have been addressed in this chapter. A more detailed investigation of the design aspects to this decoder will be given in the next chapter.

This chapter concludes the technology part of the book. The following parts are dealing with logic design aspects. Given the importance of the decoder for every crossbar circuit, the next chapter focuses on the decoder logic design methodologies. The first goal is to optimize the encoding scheme. The optimized codes are then used with the MSPT-decoder, as well as other decoder types, in order to assess the trade-offs between circuit area, yield and fabrication complexity.

References

1. Moselund KE, Bouvet D, Ben Jamaa HH, Atienza D, Leblebici Y, De Micheli G, Ionescu AM (2008) Prospects for logic-on-a-wire. Microelectron Eng 85:1406–1409
2. Auzelyte V, Solak HH, Ekinci Y, MacKenzie R, Vrs J, Olliges S, Spolenak R (2008) Large area arrays of metal nanowires. Microelectron Eng 85(5–6):1131–1134
3. Doherty L, Liu H, Milanovic V (2003) Application of MEMS technologies to nanodevices. In: ISCAS'03. Proceedings of the 2003 International Symposium on Circuits and systems, vol 3, pp III–934–III–937
4. Doherty L, Liu H, Milanovic V (2003) Application of MEMS technologies to nanodevices 3(5):III–934–III–937
5. Ng RMY, Wang T, Chan M (2007) A new approach to fabricate vertically stacked single-crystalline silicon nanowires. pp 133–136
6. Center for Micro- and Nanotechnologies (CMI) at EPFL. Available at: http://cmi.epfl.ch
7. Ben Jamaa MH, Cerofolini G, Leblebici Y, De Micheli G (2009) Nanowire Crossbar Framework Optimized for the Multi-Spacer Patterning Technique. In: Proceedings of CASES, Granoble, France
8. Wagner RS, Ellis WC (1964) Vapor-liquid-solid mechanism for single crystal growth. Appl Phys Lett 4(5):89–90
9. Holmes JD, Johnston KP, Doty RC, Korgel BA (2000) Control of thickness and orientation of solution-grown silicon nanowires. Science 287(5457):1471–1473
10. Cui Y, Duan X, Hu J, Lieber CM (2000) Doping and electrical transport in silicon nanowires. J Phys Chem B 4(22):5213–5216
11. He R, Yang P (2006) Giant piezoresistance effect in silicon nanowires. Nat Nanotechnol 1(1):42–46

12. Lauhon LJ, Gudiksen MS, Wang D, Lieber CM (2002) Epitaxial core-shell and core-multishell nanowire heterostructures. Nature 420:57–61

13. Gudiksen MS, Lauhon LJ, Wang J, Smith DC, Lieber CM (2002) Growth of nanowire superlattice structures for nanoscale photonics and electronics. Nature 415:617–620

14. Hsu J-F, Huang B-R, Huang C-S (2005) The growth of silicon nanowires using a parallel plate structure. In: The 5th IEEE Conference on Nanotechnology, vol 2, pp 605–608

15. Yang C, Zhong Z, Lieber CM (2005) Encoding electronic properties by synthesis of axial modulation-doped silicon nanowires. Science 310(5752):1304–1307

16. Hochbaum AI, Fan R, He R, Yang P (2005) Controlled growth of Si nanowire arrays for device integration. Nano Lett 5(3):457–460

17. Schmidt V, Riel H, Senz S, Karg S, Riess W, Gösele U (2006) Realization of a silicon nanowire vertical surround-gate field-effect transistor. Small 2(1):85–88

18. Hayden O, Björk M, Schmid H, Riel H, Drechsler U, Karg S, Lörtscher E, Riess W (2007) Fully depleted nanowire field-effect transistor in inversion mode. Small 3(2):230–234

19. Huang Y, Duan X, Wei Q, Lieber CM (2001) Directed assembly of one-dimensional nanostructures into functional networks. Science 291(5504):630–633

20. Moselund KE, Bouvet D, Tschuor L, Pot V, Dainesi P, Eggimann C, Thomas NL, Houdré R, Ionescu AM (2007) Cointegration of gate-all-around MOSFETs and local silicon-on-insulator optical waveguides on bulk silicon. IEEE Trans Nanotechnol 6(1):118–125

21. Lee K-N, Jung S-W, Kim W-H, Lee M-H, Shin K-S, Seong W-K (2007) Well controlled assembly of silicon nanowires by nanowire transfer method. Nanotechnology 18(44):445302 (7pp)

22. Suk SD, Lee S-Y, Kim S-M, Yoon EJ, Kim M-S, Li M, Oh CW, Yeo KH, Kim SH, Shin D-S, Lee K-H, Park HS, Han JN, Park C, Park J-B, Kim D-W, Park D, Ryu B-I (2005) High performance 5 nm radius twin silicon nanowire MOSFET (TSNWFET): fabrication on bulk Si wafer, characteristics, and reliability. In: IEEE Transactions on Nanotechnology, pp 717–720

23. Koo S-M, Fujiwara A, Han J-P, Vogel EM, Richter CA, Bonevich JE (2004) High inversion current in silicon nanowire field effect transistors. Nano Lett 4(11):2197–2201

24. Kedzierski J, Bokor J (1997) Fabrication of planar silicon nanowires on silicon-on-insulator using stress limited oxidation. J Vac Sci Technol B 15(6):2825–2828

25. Vazquez-Mena O, Villanueva G, Savu V, Sidler K, van den Boogaart MAF, Brugger J (2008) Metallic nanowires by full wafer stencil lithography. Nano Lett 8(11):3675–3682

26. Hållstedt J, Hellström PE, Zhang Z, Malm B, Edholm J, Lu J, Zhang SL, Radamson H, Östling M (2006) A robust spacer gate process for deca-nanometer high-frequency MOSFETs. Microelectron Eng 83(3):434–439

27. Choi Y-K, Lee JS, Zhu J, Somorjai GA, Lee LP, Bokor J (2003) Sublithographic nanofabrication technology for nanocatalysts and DNA chips. J Vac Sci Technol B: Microelectron Nanometer Struct 21:2951–2955

28. Cerofolini G (2007) Realistic limits to computation. II. The technological side. Appl Phys A 86(1):31–42

29. Wu W, Jung G-Y, Olynick DL, Straznicky J, Li Z, Li X, Ohlberg DAA, Chen Y, Wang S-Y, Liddle JA, Tong WM, Williams RS (2005) One-kilobit cross-bar molecular memory circuits at 30-nm half-pitch fabricated by nanoimprint lithography. Appl Phys A: Mater Sci Process 80(6):1173–1178

30. Jung GY, Johnston-Halperin E, Wu W, Yu Z, Wang SY, Tong WM, Li Z, Green JE, Sheriff BA, Boukai A, Bunimovich Y, Heath JR, Williams RS (2006) Circuit fabrication at 17 nm half-pitch by nanoimprint lithography. Nano Lett 6(3):351–354

31. Sonkusale SR, Amsinck CJ, Nackashi DP, Spigna NHD, Barlage D, Johnson M, Franzon PD (2005) Fabrication of wafer scale, aligned sub-25nm nanowire and nanowire templates using planar edge defined alternate layer process. Phys E: Low-dimensional Syst Nanostruct 28(2):107–114

32. Smith PA, Nordquist CD, Jackson TN, Mayer TS, Martin BR, Mbindyo J, Mallouk TE (2000) Electric-field assisted assembly and alignment of metallic nanowires. Appl Phys Lett 77:1399–1401

33. Duan X, Huang Y, Cui Y, Wang J, Lieber CM (2001) Indium phosphide nanowires as building blocks for nanoscale electronic and optoelectronic devices. Nature 409:66–69

34. Chen Y, Ohlberg DAA, Li X, Stewart DR, Stanley Williams R, Jeppesen JO, Nielsen KA, Stoddart JF, Olynick DL, Anderson E (2003) Nanoscale molecular-switch devices fabricated by imprint lithography. Appl Phys Lett 82:1610–1612

35. Zasadzinski JA, Viswanathan R, Madsen L, Garnaes J, Schwartz DK (1994) Langmuir-Blodgett films. Science 263(5154):1726–1733

36. Wu W, Jung G-Y, Olynick DL, Straznicky J, Li Z, Li X, Ohlberg DAA, Chen Y, Wang S-Y, Liddle JA, Tong WM, Williams RS (2005) One-kilobit cross-bar molecular memory circuits at 30-nm half-pitch fabricated by nanoimprint lithography. Appl Phys A: Mater Sci Process 80(6):1173–1178

37. Melosh NA, Boukai A, Diana F, Gerardot B, Badolato A, Petroff PM, Heath JR (2003) Ultrahigh-density nanowire lattices and circuits. Science 300(5616):112–115

38. Green JE, Wook Choi J, Boukai A, Bunimovich Y, Johnston- Halperin E, Deionno E, Luo Y, Sheriff BA, Xu K, Shik Shin Y, Tseng HR, Stoddart JF, Heath JR (2007) A 160-kilobit molecular electronic memory patterned at 10^{11} bits per square centimetre. Nature 445: 414–417

39. Luo Y, Collier CP, Jeppesen JO, Nielsen KA, DeIonno E, Ho G, Perkins J, Tseng H-R, Yamamoto T, Stoddart JF, Heath JR (2002) Two-dimensional molecular electronics circuits. J Chem Phys Phys Chem 3:519–525

40. Ho G, Heath JR, Kondratenko M, Perepichka DF, Arseneault K, Pézolet M, Bryce MR (2005) The first studies of a tetrathiafulvalenesigma- acceptor molecular rectifier. Chem—A Eur J 11(10):2914–2922

41. McCreery RL (2004) Molecular electronic junctions. Chem Mater 16(23):4477–4496

42. Ashwell GJ, Urasinska B, Tyrrell WD (2006) Molecules that mimic Schottky diodes. Phys Chem Chem Phys (Incorporating Faraday Transactions) 8:3314–3319

43. Collier CP, Mattersteig G, Wong EW, Luo Y, Beverly K, Sampaio J, Raymo FM, Stoddart JF, Heath JR (2000) A [2]catenanebased solid state electronically reconfigurable switch. Science 289:1172–1175

44. Zhang Y, Kim S, McVittie J, Jagannathan H, Ratchford J, Chidsey C, Nishi Y, Wong H-S (2007) An integrated phase change memory cell with ge nanowire diode for cross-point memory. In: IEEE Symposium on VLSI Technology, pp 98–99

45. Voutsas AT, Hatalis MK (1992) Structure of as-deposited LPCVD silicon films at low deposition temperatures and pressures. J Electrochem Soc 139(9):2659–2665

46. Voutsas AT, Hatalis MK (1993) Surface treatment effect on the grain size and surface roughness of as-deposited LPCVD polysilicon films. J Electrochem Soc 140(1):282–288

47. Voutsas AT, Hatalis MK (1993) Deposition and crystallization of a-Si low pressure chemically vapor deposited films obtained by low-temperature pyrolysis of disilane. J Electrochem Soc 140(3):871–877

48. Pott V (2008) Gate-all-around silicon nanowires for hybrid single electron transistor/CMOS applications. Ph.D. dissertation, Lausanne, 2008. Available at: http://library.ep.ch/theses/?nr=3983

49. Nakazawa K (1991) Recrystallization of amorphous silicon films deposited by low-pressure chemical vapor deposition from Si_2H_6 gas. J Appl Phys 69(3):1703–1706

50. Bergamini F, Bianconi M, Cristiani S, Gallerani L, Nubile A, Petrini S, Sugliani S (2008) Ion track formation in low temperature silicon dioxide. Nucl Instrum Methods Phys Res Sect B: Beam Interact Mater Atoms 266(10):2475–2478

51. Byon K, Tham D, Fischer JE, Johnson AT (2007) Systematic study of contact annealing: ambipolar silicon nanowire transistor with improved performance. Appl Phys Lett 90(14):143513

52. Appenzeller J, Knoch J, Tutuc E, Reuter M, Guha S (2006) Dualgate silicon nanowire transistors with nickel silicide contacts. In: International Electron Devices Meeting, pp 1–4
53. Weber WM, Geelhaar L, Graham AP, Unger E, Duesberg GS, Liebau M, Pamler W, Cheze C, Riechert H, Lugli P, Kreupl F (2006) Silicon-nanowire transistors with intruded nickel-silicide contacts. Nano Lett 6(12):2660–2666
54. Koo S-M, Edelstein MD, Li Q, Richter CA, Vogel EM (2005) Silicon nanowires as enhancement-mode Schottky barrier field-effect transistors. Nanotechnology 16(9):1482–1485
55. Ecoffey S, Mazza M, Pott V, Bouvet D, Schmid A, Leblebici Y, Declereq M, Ionescu A (2005) A new logic family based on hybrid MOSFET-polysilicon nanowires. pp 269–272
56. Hogg T, Chen Y, Kuekes P (2006) Assembling nanoscale circuits with randomized connections. IEEE Trans Nanotechnol 5(2):110–122
57. Beckman R, Johnston-Halperin E, Luo Y, Green JE, Heath JR (2005) Bridging dimensions: demultiplexing ultrahigh density nanowire circuits. Science 310(5747):465–468
58. Chua L (1971) Memristor—the missing circuit element. IEEE Trans Circuit Theory 18(5):507–519
59. Hodgkin AL, Huxley AF (1952) A quantitative description of membrane current and its application to conduction and excitation in nerve. J Physiol 117:500–544
60. Linares-Barranco B, Serrano-Gotarredona T (2009) Memristance can explain spike-time-dependent plasticity in neural synapses. In: Nature Precedings, pp 1–4
61. Smerieri A, Berzina T, Erokhin V, Fontana MP (2008) A functional polymeric material based on hybrid electrochemically controlled junctions. Mater Sci Eng: C 28(1):18–22
62. Borghetti J, Li Z, Straznicky J, Li X, Ohlberg DAA, Wu W, Stewart DR, Williams RS (2009) A hybrid nanomemristor/transistor logic circuit capable of self-programming. Proc Natl Acad Sci 106:1699–1703
63. Toumazou C, Georgiou J, Drakakis E (1998) Current-mode analogue circuit representation of Hodgkin and Huxley neuron equations. Electron Lett 34(14):1376–1377
64. Strukov DB, Snider GS, Stewart DR, Williams RS (2008) The missing memristor found. Nature 453:80–83
65. Stewart DR, Ohlberg DAA, Beck PA, Chen Y, Williams RS, Jeppesen JO, Nielsen KA, Stoddart JF (2004) Molecule independent electrical switching in Pt/organic monolayer/Ti devices. Nano Lett 4(1):133–136
66. Jo SH, Kim K-H, Lu W (2009) High-density crossbar arrays based on a Si memristive system. Nano Lett 9(2):870–874
67. Shenoy R, Gopalakrishnan K, Rettner C, Bozano L, King R, Kurdi B, Wickramasinghe H (2006) A new route to ultra-high density memory using the micro to nano addressing block (MNAB). In: VLSI Technol., pp 140–141
68. Gopalakrishnan K, Shenoy RS, Rettner C, King R, Zhang Y, Kurdi B, Bozano LD, Weslser JJ, Rothwell MB, Jurich M, Sanchez MI, Hernandez M, Rice PM, Risk WP, Wickramasinghe HK (2005) The micro to nano addressing block. In: IEEE Electron Devices Meeting, p 19.4
69. DeHon A, Lincoln P, Savage J (2003) Stochastic assembly of sublithographic nanoscale interfaces. IEEE Trans Nanotechnol 2(3):165–174
70. Ben Jamaa MH, Leblebici Y, De Micheli G (2009) Decoding nanowire arrays fabricated with the multi-spacer patterning technique. In: Design Automation Conference (DAC), San Francisco, California, USA
71. International technology roadmap for semiconductors (ITRS) (2007) http://www.itrs.net/reports.html. Tech. Rep., 2007

Chapter 3
Decoder Logic Design

Among the different emerging technologies surveyed previously in Chap. 1, crossbars are a very promising approach to integrate silicon nanowires and molecular switches into functional circuits. Chapter 2 proposed a fabrication framework for crossbars, the multi-spacer patterning technique, which has the advantage of being CMOS compatible and using only photolithography steps. Other approaches to fabricate crossbars reported in literature include nanomolds and self-assembly.

The fundamental difference between crossbars lies in the way nanowires are fabricated. As explained in Chap. 2, nanowires can be fabricated with top-down and bottom-up techniques. Top-down techniques generally use a certain kind of patterning, such as photolithography, while in bottom-up approaches nanowires are grown from a seed and need to be transferred onto the functional substrate. This transfer is generally carried out by means of fluid-assisted deposition and self-assembly of the nanowires.

The motivation towards the crossbar architecture has many reasons. First, the ability to manipulate and place single nanowires is limited. Some methods exist, such us the use of *atomic force microscope* (*AFM*) to manipulate single nanowires, but they cannot be deployed in mass production of VLSI systems. Second, the nanowire technology is immature and consequently unreliable. Then, it is highly desirable to organize the circuits into fault-tolerant structures with a large regularity and redundancy levels. Finally, crossbars can implement some circuit families in a more area-efficient way than CMOS, such as *random access memory* (*RAM*), *look-up tables* (*LUT*) or two-level logic circuits. They promise to reach the ultimate physical limit of memory and computation with the actual electron-based VLSI paradigm.

The ability of crossbars to perform computation and memory in an area-efficient way with a low reliability level motivates their implementation with a hybrid architecture including both CMOS and crossbar parts. The CMOS part is larger in area than the crossbar part, while it is more reliable. The signals coming from the CMOS part are routed to every crossbar. Hence, it is necessary to link the CMOS part, defined with standard photolithography, to the crossbar part, defined on the

M. H. Ben Jamaa, *Regular Nanofabrics in Emerging Technologies*,
Lecture Notes in Electrical Engineering, 82, DOI: 10.1007/978-94-007-0650-7_3,
© Springer Science+Business Media B.V. 2011

sub-photolithographic scale. The interface between these two parts is the decoder, which has been the core of many research works, and which represents to topic of this chapter.

The decoder is a critical part that highly depends on the underlying nanowire technology [1–5]. The design of the decoder needs to carefully address the variability of the nanowires. In previous approaches, no decoder design technique has been suggested for the MSPT technology; and for other technologies, only binary codes have been used. The scope of this chapter is to enlarge the design space for the decoder, by developing new code families that can enhance the decoder fault-tolerance, save area and reduce the fabrication complexity. The introduced novel codes are based on a *multi-valued logic* (*MVL*), and some of them are derived from well known codes by arranging their elements in an optimized way.

Parts of this chapter have been published in [6, 7]. It is organized as follows. The crossbar architecture is first introduced, showing the different parts of the circuit and highlighting the decoder part. Then, previously proposed decoder and encoding schemes are surveyed. The construction of new families of MVL codes is then introduced, and the ability of these codes to uniquely address nanowires and improve the decoder defect-awareness while saving area is demonstrated. Thereafter, the MSPT decoder methodology is presented and its defect tolerance and technology complexity are optimized by a set of codes, some of which are specifically developed for this purpose. Finally, the chapter is concluded by a discussion of the obtained results and an assessment of the contributions of this part of the work.

3.1 Crossbar Architecture

The baseline organization of a nanowire crossbar circuit is depicted in Fig 3.1a. An arrangement of two orthogonal layers of parallel nanowires defines a regular grid of intersections called crosspoints. The separation between the two layers can be filled with a phase change material or molecular switches at the crosspoints. Information storage, interconnection or computation can be performed with these crosspoints [8, 9]. A set of contact groups is defined on top of the nanowires. Every contact group makes an ohmic contact to a corresponding distinct set of nanowires, which represents the smallest set of nanowires that can be contacted by the lithographically defined lines, called *mesowires* (*MWs*).

This configuration bridges every set of nanowires within a contact group to the outer CMOS circuit. In order to *fully bridge the scales* and make every nanowire within this set uniquely addressable by the outer circuit, a decoder is needed. It is formed by a series of transistors along the nanowire body, controlled by the mesowires and having different threshold voltages V_T (Fig. 3.1b). The distributions of V_T's is called the *nanowirepattern*. Depending on this pattern and the pattern of applied voltages in the decoder (V_A's), one single nanowire in the array can be made conductive (Fig. 3.1c). In this case, this nanowire is said to be addressed by the applied voltage pattern.

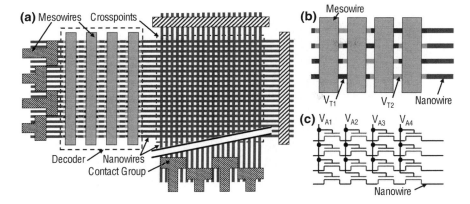

Fig. 3.1 Baseline organization of a crossbar circuit and its decoder. **a** Architecture of a crossbar circuit. **b** Decoder layout. **c** Decoder circuit design

It is possible to think of replacing each transistor at the diagonal crosspoints by an ohmic contact and to eliminate all other transistors; thus, mapping each horizontal wire onto a vertical. However, this method is technologically difficult, because the nanowire pitch is defined below the photolithographic limits. Moreover, when the number of horizontal wires (i.e., the NWs) becomes large and their size small, while that of the addressing wires (i.e., the MWs) remains at the lithographic scale, the decoder size increases and it becomes larger than the size of the crossbar part of the circuit. Then, the area gain given by the compactness of the crossbar part can be canceled by the large area of the MWs in the decoder part. By encoding the NWs, however, the number of MWs scales only logarithmically with the number of NWs. Table 3.1 illustrates the area of crossbars realized with different lithography pitches (L_l) and sub-lithographic or nanoscale pitches (L_n), and for different crossbar densities (D). The number of NWs N was derived from the crossbar density, whereas the number of MWs M was scaled logarithmically with N.

3.2 Decoder and Encoding Types

The decoder is the fundamental element of the crossbar circuit that bridges the scales. Its design highly depends on the underlying nanowire fabrication technology. Yield, area and complexity of the decoder are highly depending on the

Table 3.1 Area in μm^2 of a crossbar circuit for different technologies and crossbar densities

D (kB)	N	M	$L_l = 65$ nm		$L_l = 32$ nm	
			$L_n = 10$ nm	$L_n = 5$ nm	$L_n = 10$ nm	$L_n = 5$ nm
32	512	9	129	38	117	32
64	724	10	247	71	228	62

encoding scheme as well, making the encoding scheme an important system-level design parameter. In this section, previous design and fabrication approaches of nanowire decoders and usually used encoding schemes are surveyed.

3.2.1 Decoder Design and Fabrication

Even though the structure of the decoder circuit is simple, its reliable fabrication and design are challenging. The need to use different transistors necessitates different doping levels in specific regions on the nanowires whose location cannot be controlled precisely because the nanowire scale is below the lithographic limit. Thus, nanowires that are already doped during the fabrication process may simplify the task. When it comes to the decoder, it is fundamental to consider the nanowire fabrication technique and distinguish between differentiated and undifferentiated nanowires. Differentiated nanowires are those having a certain doping profile; they are generally fabricated in a bottom-up approach and the doping profile is defined during the nanowire growth. Undifferentiated nanowires have no specific doping profile; they are identical to each other, and they are generally fabricated in a top-down approach. Today, different techniques have been proposed to fabricate and design the decoder for both differentiated and undifferentiated nanowires.

3.2.1.1 Decoders for Differentiated Nanowires

Differentiated NWs have an axial or a radial doping profile which was defined during the NW growth process. An axial decoder was presented in [1], in which the distribution of the V_T's is fully random. The NWs are dispersed parallel to each other and they are addressable when they have different V_T patterns. The probability that their addresses are different may be increased by increasing the number of addressing wires. On the other hand, the radial decoder [2] relies on NWs with several radial doping shells. The remaining shells after a sequence of etchings depends on the etching order in every region. The suite of shells along the NW after all etching steps defines the NW patterns. While both axial and radial decoder give the same estimate of the number of MWs needed to address the available NWs; the radial decoder has the advantage of being less sensitive to misalignment of NWs.

Assuming that the doping regions have the same width but different patterns, then the NWs laid out parallel to each other can be addressed by the crossing perpendicular MWs. If each NW is chosen with a known probability of having a given code, then the probability that it has a unique code (i.e., sequence of doping regions) increases with the number of codes. Depending on the NW code, a certain sub-set of MWs will prohibit the conduction in almost all other NWs except the one with this given code. To address N NWs, M MWs are needed, $M = \lfloor 2.2 \cdot \log_2(N) \rfloor + 11$. With these dimensions, the axial decoder works properly, i.e., it uniquely addresses the nanowires, with a probability greater than 99%, but it does not guarantee that the doping regions will lie directly under the MWs. The misalignment induces a slight decrease in the effective number of addressed NWs [1].

However, in the radial decoder [2], the radially doped NWs are self-aligned, and then exposed to different etchants in a certain order. The horizontal position of the etched regions is precisely defined by lithography, and the MWs are laid out at these positions. The remaining shell in each region depends on the etching order and the suite of shells along the NW after all etching steps defines the NW code. Therefore, the etching process takes care of precisely defining the sub-lithographic vertical dimensions of the contact regions. The radial decoder is technologically different from the axial decoder, but gives the same estimate of the number of MWs needed to address the available NWs, and is less sensitive to misalignment of NWs.

3.2.1.2 Decoders for Undifferentiated Nanowires

On the other hand, for undifferentiated nanowires, namely those fabricated in a top-down process, a mask-based decoder was presented in [10] and its ability to control undifferentiated NWs was proven. The MWs are separated from the NWs by a non-uniform oxide layer: in some locations a high-κ is used, in the others a low-κ dielectric. The high-κ dielectric amplifies the electric field generated by the MWs relatively to the low-κ dielectric. Consequently, the field effect control by the MWs happens only at the NW regions lying under the high-κ dielectric. The oxide mask is lithographically defined; making the decoder depending on the lithography limits. In order to address N nanowires, the mask-based decoder necessitates the use of $M = 2 \cdot \log_2(N) + \epsilon$ mesowires, with ϵ a small constant ≥ 1, which depends on the fabrication technique and the degree of redundancy to be achieved.

For undifferentiated NWs, a random contact decoder has been presented in [3, 4]. Unlike the other decoders for which the NW codes are among a known set of codes, the connections established between MWs and NWs for this decoder are fully random. It results from a deposition of gold particle onto the NWs, where the only controlled parameter is the density of particles. In order to control each of the N NWs uniquely with a high probability, $M = 4.8 \cdot \log_2(N) + C$ mesowires are needed, with C a large constant that depend on the design parameters.

Recently, a decoder called Micro to Nano Addressing Block, MNAB, has been presented in [5] for undifferentiated NWs. Due to its analogue working principle, it needs only two MWs to address any number of NWs within a certain range depending on the technology used. The two MWs are laid out parallel to all the NWs and create an electric field in the NW array by means of the voltages applied at them. Depending on the voltage applied at each MW, the minimum of the electric field can be set at any NW which will conduct; the resistances of all other NWs will highly increase.

3.2.2 Encoding Schemes

Various code types have been investigated for decoding nanowire arrays, all of them are defined with binary logic. The binary hot code, or simply, *hot code* (*HC*), is defined with two parameters (M, k), and it spans the space of code words with the length M having k occurrences of the bit '1' and $(M - k)$ occurrences of the bit '0' in every code word $(k \leq M)$. It is also known as the k-out-of-M code; which was first used as a defect/tolerant encoding scheme [11]. For instance, the code words 001111 and 010111 belong to the same hot code space with $(M, k) = (6, 4)$.

The binary tree code, or simply *tree code* (*TC*), with the length M is a 2-to-2^M encoder representing the 2^M binary numbers 0...0 to 1...1. For instance, 000, 001 and 101 are elements of the TC space with the length $M = 3$. However, in order for tree codes to uniquely address nanowires, it is necessary to make them *reflexive*, i.e., to append to every code word its n-complement. For instance, 000, 001 and 101 become respectively 000111, 001110 and 101010. These reflected (binary) tree codes are called, *binary reflexive codes BRC*.

Gray codes (GC) are known for their interesting properties that enhance the fault tolerance in many applications [12]. However, there has been no attempt to use Gray codes to design nanowire decoders. Gray codes are an arrangement of tree codes where successive code words differ in only one digit. For instance, the sequence of code words 010 $\Rightarrow$ 010 taken from the binary tree code is not allowed in Gray codes, because it has two transitions, at the second and the third digits. In a GC, 010 can be only followed by 110, 000 or 011, in order to keep the transition count qual to one.

There are various types of Gray codes. *The* Gray code (*GC*) means in this work the first patented Gray code [13]. In addition, we consider also the *balanced Gray code* (*BGC*) [14], where digit changes are distributed as equally as possible among all digit positions. In this work, we assume that the digit change is 2 or less whenever we refer to the BGC. The way such a code can be derived was explained in [14].

The Gray code and its various versions are special arrangements of the tree code. Thus, in order to make nanowires addressable by these codes, we need to use their reflected form by appending to every code word its complement. We will omit to specify that the used TC, GC and BGC are reflected but we will assume it since all the codes cited in the work are used to address nanowires.

3.3 Multi-Valued Logic Encoding

Previously explored nanowire encoding schemes, i.e., codes, are binary. The code length impacts the decoder size and the overall crossbar area. It is therefore interesting to investigate the benefits of reducing the code length by using MVL codes. The generalization of the usual codes to MVL produces novel code families that have not been explored before. In this section, the construction rules for new code families are presented. Defects that can affect them are modeled. Then, the fault-tolerance of the considered codes and their impact on the crossbar circuit in terms of reliability and area are investigated.

3.3.1 Circuit Design with Multi-Valued Logic

The research areas for multi-valued logic can be summarized in three categories: MV algebra, MV semiconductor circuits and MV network synthesis. A review of the background of MV algebra was presented in [15] and [16]. The implementation of algebraic notions into real circuits was motivated from one side by the exponential growth of interconnects in digital circuits and their limited scaling abilities [15], and from the other side by the need for higher density of information storage. The use of MVL encoding of data reduces the area needed for MVL buses and memories.

Many circuit design techniques were used to implement MVL: different current-mode MVL circuits were reviewed in [17]; while in [18] a voltage-mode MVL full adder was demonstrated. A charge- and voltage-based approach for MVL Flash memories was described in [19] and MVL SRAM memory and logic blocks fabricated with the same technology were introduced in [20].

Design tools for mapping MVL functions onto FPGA systems were presented in [21]. Optimized design methodologies for MVL PLA were introduced in [22] and [23] while considering the MVL encoding problem of inputs and outputs and the benefits of encoding in terms of number of products in the minimized function. Simplification and minimization of EXOR-sum-of-product expressions for MVL functions were presented in [24] and [25] respectively. The use of MV functional decomposition algorithms based on MV decision diagrams was investigated in [26] for logic synthesis. Efficient minimization of MVL networks with $don't - cares$ was presented in [27] as a way to implement MVL hardware, and also as an optimization opportunity for binary functions at the MV stage, which cannot be discovered in the binary domain.

3.3.2 Semantic of Multi-Valued Logic Addressing

In this section, we generalize the notion of encoding to multiple-valued bits by first defining some basic relations needed to identify possible codes. Some basic concepts used in encoding theory are generalized from the binary definitions stated in [28] to the multiple-valued logic. The matching between a code and its pattern corresponds here to conduction. Before introducing the impact of defects, we consider the code (Ω) and pattern (A) spaces to be identical, realizing a 1-to-1 mapping between each other. Algebraic operations are performed as defined in the ring of integers.

Definition 1 A multiple-valued *pattern* **a**, or simply a pattern **a**, is a suite of M digits a_i, in the n-valued base $\mathbb{B}$; i.e., $\mathbf{a} = (a_0, \ldots, a_{M-1}) \in \mathbb{B}^M, \mathbb{B} = \{0 \ldots n-1\}$.

A pattern represents a serial connection of M transistors in the silicon nanowire core; each digit a_i of the code word represents a threshold voltage $V_{T,i}$, with the convention $a_i < a_j \Leftrightarrow V_{T,i} < V_{T,j} \forall i, j = 0 \ldots M-1$. An analogue equivalence holds for $a_i = a_j$ and consequently for $a_i > a_j$. This convention is equivalent to discretizing the M values of V_T and ordering them in an increasing order. In Fig. 3.2a, b we illustrated the pattern 002120 representing the V_T sequence (0.2 V, 0.2 V, 0.6 V, 0.4 V, 0.6 V, 0.2 V).

Definition 2 A multiple-valued *code word* **c**, or simply a code word **c**, is, similarly to a pattern, a suite of M digits c_i, in the n-valued base $\mathbb{B} = \{0, \ldots, n-1\}$; i.e., $\mathbf{c} = (c_0, \ldots, c_{M-1}) \in \mathbb{B}^M$.

A code word represents the suite of applied voltages V_A at the M mesowires. These are defined such that every $V_{A,i}$ is slightly higher than $V_{T,i}$, and lower than $V_{T,i+1}$. Hence, a similar convention holds for the order of $V_{A,i}$ with respect to that of c_i. In Fig. 3.2c, d we illustrated the code word 202111 representing the sequence of applied voltages (0.7 V, 0.3 V, 0.7 V, 0.5 V, 0.5 V, 0.5 V).

Definition 3 A *complement* of digit x_i in a code word or pattern **x** is defined as: $\mathsf{NOT}(x_i) = \overline{x_i} = (n-1) - x_i$. The operator NOT can be generalized to the vector **x**, acting on each component as defined above. Notice that $\mathsf{NOT}(\mathsf{NOT}(\mathbf{x})) = \mathbf{x}$.

Definition 4 A pattern **a** is *covered* by a code word **c** if and only if the following relation holds: $\forall i = 0 \ldots M-1, c_i \geq a_i$. By using the sigmoid function

$$\sigma(x) = \begin{cases} 0 & x \leq 0 \\ 1 & x > 0 \end{cases}$$

Fig. 3.2 Mapping of threshold and applied voltages onto discretized values. **a** Pattern 002120 and its V_T sequence. **b** Discretization of V_T values. **c** Code word 202111 and its V_A sequence. **d** Discretization of V_A values

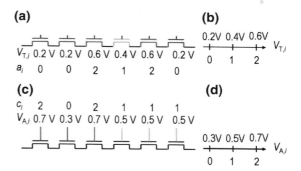

generalized to vectors: $\sigma(\mathbf{x}) = (\sigma(x_0), \ldots, \sigma(x_{M-1}))$, the definition above becomes: $\mathbf{a}$ is covered by $\mathbf{c} \Leftrightarrow \|\sigma(\mathbf{a} - \mathbf{c})\| = 0$. Alternatively, we can define the order relations on vectors $\mathbf{c}$ and $\mathbf{a}$:

$$\mathbf{c} < \mathbf{a} \Leftrightarrow \forall i, \quad c_i < a_i$$
$$\mathbf{c} > \mathbf{a} \Leftrightarrow \forall i, \quad c_i > a_i$$

The relation becomes relaxed (i.e., $\leq$ or $\geq$) if exists i such that $c_i = a_i$. Then, a pattern $\mathbf{a}$ is covered by a code word $\mathbf{c}$ if and only if $\mathbf{a} \leq \mathbf{c}$. The same definition for covering can be generalized to two patterns or two code words.

Covering a given pattern with a certain code is equivalent to applying a suite of gate voltages making every transistor conductive. Then, the nanowire is conducting and we say that it is controlled by the given sequence of gate voltages. Figure 3.3a illustrates the case in which the code covers the pattern and the nanowire is conducting, while Fig. 3.3b illustrates the opposite case.

Definition 5 A pattern $\mathbf{a}$ *implies* a pattern $\mathbf{b}$ if and only if $\|\sigma(\mathbf{b} - \mathbf{a})\| = 0$; i.e., $\mathbf{b}$ is covered by $\mathbf{a}$. We note this as follows: $\mathbf{a} \Rightarrow \mathbf{b}$. Since a 1-to-1 mapping between the patterns and codes was assumed, we generalize this definition to code words: $(\mathbf{c}^a \Rightarrow \mathbf{c}^b) \Leftrightarrow \|\sigma(\mathbf{c}^b - \mathbf{c}^a)\| = 0$; i.e., $\mathbf{c}^b$ is covered by $\mathbf{c}^a$.

This means that if a nanowire with the pattern $\mathbf{a}$ corresponding to the code $\mathbf{c}^a$ is covered by a code $\mathbf{c}^*$, then the nanowire with the pattern $\mathbf{b}$ corresponding to the code $\mathbf{c}^b$ is also covered by the same code $\mathbf{c}^*$. Applying the voltage suite $\mathbf{c}^*$ will result in turning on the nanowires with either pattern (see Fig. 3.4).

Definition 6 The code words $\mathbf{c}^a$ and $\mathbf{c}^b$ are *independently covered* if and only if $\mathbf{c}^a$ does not imply $\mathbf{c}^b$ and $\mathbf{c}^b$ does not imply $\mathbf{c}^a$.

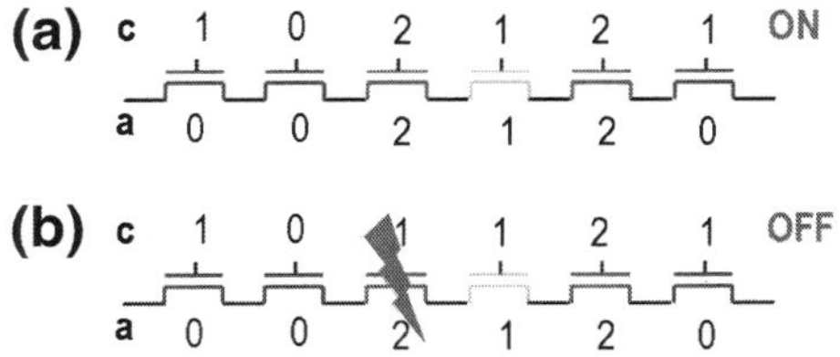

Fig. 3.3 Example of conducting and non-conducting nanowires. **a** Conducting nanowire (code covers pattern). **b** Non-conducting nanowire (code does not cover pattern)

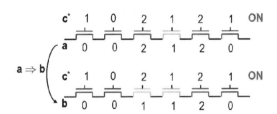

Fig. 3.4 Example of implication between patterns: $\mathbf{c}^*$ covers $\mathbf{a}$; since $\mathbf{a} \Rightarrow \mathbf{b}$, then $\mathbf{c}^*$ covers also $\mathbf{b}$

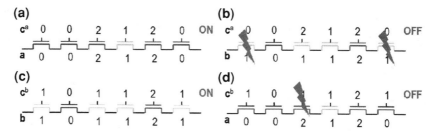

Fig. 3.5 Example of independent covering: code words $\mathbf{c}^a$ and $\mathbf{c}^b$ are independently covered

This definition means that there exists a voltage suite that turns on the nanowire with the pattern **a** corresponding to $\mathbf{c}^a$ but not that with the pattern **b** corresponding to $\mathbf{c}^b$ (see Fig. 3.5a, b). Reciprocally, there exists a second voltage pattern that turns on the nanowire with the pattern **b** corresponding to $\mathbf{c}^b$ but not that with the pattern **a** corresponding to $\mathbf{c}^a$ (see Fig. 3.5c, d).

Definition 7 The code word $\mathbf{c}^a$ belonging to the set Ω is *addressable* if and only if it does not imply any other code in $\Omega\backslash\{\mathbf{c}^a\}$. We define the set Ω to be addressable if and only if every code word in Ω is addressable.

Assuming that there is a 1-to-1 mapping between the code space Ω and the pattern space A, then saying that a code $\mathbf{c}^a$ implies no other code in $\Omega\backslash\{\mathbf{c}^a\}$ is equivalent to saying that it covers only the pattern **a** but no other pattern in $A\backslash\{\mathbf{a}\}$. Thus, there exists a voltage sequence that activates only the nanowire with the pattern **a** but no other nanowire having its pattern in $A\backslash\{\mathbf{a}\}$.

Proposition 1 *A set Ω of code words is addressable if and only if every code word in Ω is independently covered with respect to any other code word in Ω.*

Proof This follows directly from Definitions 6 and 7.   □

Consequently, an admissible set of applied voltages that uniquely addresses each nanowire corresponds to the set of code words Ω that independently covers every pattern in A. This set of patterns can be simply taken as Ω itself, if Ω is addressable.

3.3.3 Code Construction

3.3.3.1 Hot Encoding

In binary logic, the (k, M) hot code space is defined as the set of code words with the length M having k occurrences of the bit '1' and $(M - k)$ occurrences of the bit

'0' in every code word ($k \leq M$). It is also known as the k-out-of-M code; which was first used as a defect tolerant encoding scheme [11]. This definition can be generalized to the n-valued logic. We first define $\mathbf{k}$ as an n-dimensional vector $(k_0, \ldots, k_{n-1})$, such that $\sum_i k_i = M$. Then, the multi-valued $(\mathbf{k}, M)$-hot encoding is defined as the set of all code words having the length M such that each k_i represents the occurrence of the digit $i, i = 0, \ldots, n - 1$. We consider for instance the ternary logic ($n = 3$), and we set $\mathbf{k} = (4, 3, 1)$ and $M = 8$. Then, every code word in the considered $(\mathbf{k}, M)$-hot space contains $4\times$ the digit '0', $3\times$ the digit '1' and $1\times$ the digit '2'. The considered code space includes for instance the code words 00001112 and 00210110.

Proposition 2 *The code space defined by a multi-valued $(\mathbf{k}, M)$-hot encoding is addressable.*

Proof Consider two different code words c^a and c^b in the code space defined by the $(\mathbf{k}, M)$-hot encoding. Both codes are identical except at P different digits lying at the positions $p_0, \ldots, p_{P-1}$. c^b is obtained by a permutation of $\{\mathbf{c}^a_{p_0}, \ldots, \mathbf{c}^a_{p_{P-1}}\}$. Hence, there is at least one position p_i for which $\mathbf{c}^a_{p_i} > \mathbf{c}^b_{p_i}$ holds and at least one position p_j for which $\mathbf{c}^a_{p_j} < \mathbf{c}^b_{p_j}$ holds. This proves that $\|\sigma(\mathbf{c}^b - \mathbf{c}^a)\| \neq 0$ and $\|\sigma(\mathbf{c}^a - \mathbf{c}^b)\| \neq 0$ and that every two code words are independently covered. Then, Proposition 1 states that the whole code space is addressable. □

Example 1 For instance the code words $\mathbf{a} = 00001112$ and $\mathbf{b} = 00210110$ differ at the 3rd, 4th, 5th and last digits, in positions $\{2, 3, 4, 7\}$. The third digit of $\mathbf{a}$ is smaller than the third digit of $\mathbf{b}$, which proves that $\mathbf{b} \not\leq \mathbf{a}$; and the fifth digit of $\mathbf{a}$ is bigger than the fifth digit of $\mathbf{b}$; which proves that $\mathbf{a} \not\leq \mathbf{b}$. Thus, both codes are independently covered. □

Proposition 3 *The size of the code space defined by a multi-valued $(\mathbf{k}, M)$-hot encoding is maximal for $k_i = M/n \, \forall \, i = 0, \ldots, (n - 1)$. The size of the maximal-sized space is asymptotically proportional to $\propto n^M / M^{(n-1)/2}$ for a given n.*

Proof The number of code words is given by $M!/(\prod k_i!)$; which can be maximized by using the Gamma function $\Gamma(k_i + 1) = k_i!$. The Stirling formula yields the asymptotic space size for large M. □

In this work, it is implicitly meant that the $(\mathbf{k}, M)$-hot code with the maximal-sized space are used, even if just $(\mathbf{k}, M)$-hot code are mentioned.

3.3.3.2 N-ary Reflexive Code

The binary tree code with the length M is a 2-to-2^M encoder representing the 2^M binary numbers $0 \ldots 0$ to $1 \ldots 1$. Similarly, a n-ary tree code with the length M is defined as the set of n^M numbers ranging from $0 \cdots 0$ to $(n - 1) \cdots (n - 1)$. For instance: the ternary ($n = 3$) tree code with the length $M = 4$ includes all ternary logic numbers ranging between 0000 and 2222. As one can easily see, some code words imply many others from the same space: for instance 2222 implies all other codes. It is possible to prevent the inclusive character of the n-ary tree code by attaching the complement of the code word (i.e., 2222 becomes 22220000). The as-constructed code is the *N-ary Reflexive Code* (NRC).

Proposition 4 *The code space defined by the NRC is addressable.*

Proof The first (non-reflected) halves of any two code words $\mathbf{c}^a$ and $\mathbf{c}^b$ having the total length M (M is even) differ by at least one digit at say position i ($i = 0 \ldots (M/2 - 1)$). Let $\mathbf{c}^a$ be the code word such that $c_i^a < c_i^b$. The reflection implies that $c_{M/2+i}^a > c_{M/2+i}^b$. This proves that $\|\sigma(\mathbf{c}^b - \mathbf{c}^a)\| \neq 0$ and $\|\sigma(\mathbf{c}^a - \mathbf{c}^b)\| \neq 0$ and that $\mathbf{c}^a$ and $\mathbf{c}^b$ are independently covered. Then, Proposition 1 states that the whole code space is addressable.   □

Example 2 We consider for instance the ternary ($n = 3$) reflexive code words with the length $M = 8$: $\mathbf{c}^a = 22220000$ and $\mathbf{c}^b = 00122210$. The codes differ at some positions; we consider for instance the first one. The first digit of $\mathbf{c}^a$ ('2') is bigger than the first digit of $\mathbf{c}^b$ ('0'), proving that $\mathbf{c}^a \nleq \mathbf{c}^b$. The complement of the first digit is the fifth one (since $M = 8$). Because of the complementarity of the values of the digits, the fifth digit of $\mathbf{c}^a$ ('0') is smaller than the fifth digit of $\mathbf{c}^b$ ('2'), proving that $\mathbf{c}^b \nleq \mathbf{c}^a$. Consequently, $\mathbf{c}^a$ and $\mathbf{c}^b$ are independently covered.   □

In a similar way, the reflection principle works for any other code (e.g., Hamming code), making the whole code space addressable. However, in return it doubles the code length. Although the binary hot code is denser than the binary reflexive code, this statement holds for the multi-valued logic only if the codes are defect-free. This aspect is analyzed in the following sections.

3.3.4 Defect Models

The control of the silicon nanowires is based on the modulation of the threshold voltage of the control transistors. The proposed encoding schemes impose a distribution of the applied control voltages between the successive threshold voltages. The main issue with the threshold voltage is its variability and process-dependency: several aspects of insufficient controllability of the technology are expressed on the device level as a random variation of the threshold voltage (e.g., variation of the doping level, oxide thickness, mechanical stress of the nanowires, etc). The independent nature of these random effects and their superposition generally justify the assumption of a normal distribution of the threshold voltage (see Fig. 3.6). The MVL encoding only depends on the value of V_T's; however other random defects, such as the nanowire breakage and the nanowire-to-metal contact quality impact the array yield (but not the addressable code space size). These random effects were modeled by a statistical factor explained in Sect. 3.3.7.

Fig. 3.6 Coding defects induced by V_T variability

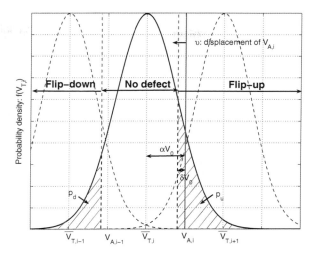

Many random errors can affect the molecular switches (bad switching, stack-at-defects,...) are not modeled, because they do not affect the decoder yield; so that their impact on the overall crossbar yield can be modeled by including a multiplication factor that reflects the statistical defects affecting the molecular switches.

3.3.4.1 Basic Error Model

Figure 3.6 illustrates the main assumptions for basic error models. We assume that the threshold voltages $V_{T,i}$ are equidistant; i.e., $V_{T,i+1} - V_{T,i} = 2\alpha V_0$, V_0 being a given scaling voltage and α is given by the technology. The applied voltages $V_{A,i}$ are set between every two successive threshold voltages $V_{T,i}$ and $V_{T,i+1}$, not necessarily in the middle, rather shifted by vV_0 towards $V_{T,i}$; where v is a design parameter.

If the variability of $V_{T,i}$ is high or the spacing between two successive $V_{T,i}$'s is low due to the large number of doping levels, then $V_{T,i}$ may exceed a voltage $V_{X,i}$ given by $V_{A,i} - \delta \cdot V_0$; where δ will be derived in the following. While V_T increases, the sensed current while a_i is applied on the digit c_i decreases ($a_i = c_i$) and the sensed current while $a_i + 1$ is applied on the same digit increases. The voltage $V_{X,i}$ is defined as the gate voltage which results in the decrease of the sensed current for a_i by the factor q from its value at $\overline{V}_{T,i}$. The higher q, the more accurate is the sensing. Thus, q is also considered as a design parameter. Assuming that the transistors are saturated, then the current in the saturation region is proportional to $(V_{A,i} - V_T)^2$, where V_T is the actual threshold voltage. Consequently, the following condition on V_X must hold: $(V_{A,i} - \overline{V}_{T,i})^2/(V_{A,i} - V_{X,i})^2 = q$; which gives: $\delta = (\alpha + v)/\sqrt{q}$ for long channel transistors.[1] This fixes the values of $V_{X,i}$; when $V_{T,i}$ exceeds $V_{X,i}$, the digit a_i acts as $a_i + 1$; its address becomes $c_i + 1$ and we call this case the *flip-up defect*.

Now, consider the case when $V_{T,i}$ falls below $V_{A,i-1} - \delta \cdot V_0 = V_{X,i-1}$, then the current flowing while $a_i - 1$ is applied is not ~ 0 anymore, and always greater than q times the current flowing while $c_i - 2$ is applied. Then, a_i is implied by c_i and $c_i - 1$ but not by $c_i - 2$; its address is $c_i - 1$ which means that a_i acts as $a_i - 1$; this case is called the *flip-down defect*. The probabilities of flip-ups and -downs are given by the following expressions, which are independent of i. Here, f_i is the probability density function of $V_{T,i}$:

$$p_u = \int_{V_{X,i}}^{\infty} f_i(x)dx \quad p_d = \int_{-\infty}^{V_{X,i-1}} f_i(x)dx$$

When $V_{T,i}$ falls within the range between the threshold values for flip-up and flip-down defects, the digit is correctly interpreted. We notice that the flip-down error never happens at digits having the smallest value, 0, since the corresponding $\overline{V}_{T,i}$ is by definition smaller than the smallest $V_{A,i}$ available. For the same reason, the flip-up error never happens at the digits having the biggest value, $n - 1$. In order to study the size of the addressable code space, we consider flip-up and -down errors in the code space instead of flip-up and -down defects at the nano-wires, since both considerations are equivalent.

3.3.4.2 Overall Impact of Variability

If V_T varies within a small range close to its mean value, then the pattern does not change, since the nanowire still conducts under the same conditions. Then, a 1-to-1 mapping between the code and the pattern space holds, which is shown in Fig. 3.7 for a ternary hot code with $M = 3$. On the contrary, if the V_T variation is large, then some digits may be shifted up or down, as explained above. When a pattern has a sequence of errors, it can be either covered by one or more codes or it can be uncovered. When we consider the codes, some of them cover one or more patterns and some cover no pattern under the error assumptions. The following example explains this conjecture:

Example 3 Figure 3.7b illustrates the digit shift at some patterns. We notice that the first pattern 022 (which underwent a defect) is not covered by any code anymore. Thus its nanowire cannot be addressed. All the other patterns are covered at least by one code. Two categories among these covered patterns can be

---

[1] If we consider short channel transistors, then the saturation current is proportional to $(V_{A,i} - V_T)$ and $\delta = (\alpha + v)/q$

Fig. 3.7 Mapping of the code space onto the pattern space. **a** Mapping in the defect-free case. **b** Mapping and in the case of defects

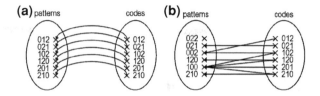

distinguished. On the one hand, the fourth pattern 120 is covered by the fourth code, which in turn covers another pattern (the fifth). Thus, by activating the fourth nanowire, the required control voltages activate either the fourth and fifth nano-wire. Consequently, the fourth nanowire cannot be addressed uniquely. This case represents the patterns covered only by codes covering more than a single pattern. On the other hand, the complementary case is illustrated by the fifth pattern 100, which is covered by many codes. However, one of these codes (201) covers no other pattern except the considered one. Thus, it is possible to uniquely activate the fifth nanowire by applying the voltage sequence corresponding to the code 201.   □

The examples shown in Fig. 3.7 demonstrate that a pattern undergoing defects can be either *i*) not covered by any valid code word, in which case the nanowire cannot be identified as addressable and the pattern is useless; or *ii*) covered by at least one valid code word. In the second case, if two patterns or more are covered by the same code word, then this code word cannot be used because more than one nanowire would have the same address. Thus, in the second case, the pattern is only useful if at least one code word covering it covers no other pattern, insuring that the covered pattern can be addressable.

Assuming that in average every code word covers v patterns when errors happen, let p_I be the probability that a pattern becomes uncovered, and p_U the probability that a code word covers a unique pattern ($\overline{p}_U = 1 - p_U$). Let $|\Omega|$ be the original size of the code space and $|\Omega'|$ the size after errors happen. The set Ω' contains the useful addresses under defect conditions, i.e., those that address unique nanowires even though the nanowires are undergoing defects. The size of Ω' indicates the number of nanowires that remain useful under high variability conditions. Then:

$$|\Omega'| = |\Omega| \cdot (1 - p_I)(1 - \overline{p}_U^v) \tag{3.1}$$

In order to assess $|\Omega'|$, we model multi-digit errors in the following sections. Then, we analytically derive p_I and p_U, and we estimate v as a fit parameter from Monte Carlo simulations.

3.3.5 Errors in the k-Hot Code Space

3.3.5.1 Error Types

Ω refers in the following to the code space of the maximal-sized multi-valued $(\mathbf{k}, M)$ hot encoding in the base $\mathbb{B} = \{0, \ldots, n-1\}$ with $\mathbf{k} = (k, \ldots, k)$ and $M = k \cdot n$. We consider a code word $\mathbf{c}$ in Ω undergoing a series of single digit errors. The multi-digit error is described by the following vector $\mathbf{d} = (\mathbf{d}_0, \ldots, \mathbf{d}_{n-1})$, where each $\mathbf{d}_i$ represents a pair of integers (d_i^u, d_i^d) expressing the number of flip-ups and flip-downs occurring in each digit group having the value i. Since no flip-down occurs at digits with value 0 and no flip-up occurs at digits with value $(n-1)$, then we impose $d_0^d = 0$ and $d_{n-1}^u = 0$. For instance, we consider the ternary hot code 010221 undergoing a flip-up defect at the third digit, turning it into 011221. One digit '0' flipped up, while no flip-error happened at the digits '1' and '2'. This error is represented as $\mathbf{d} = \left(\begin{pmatrix} 1 \\ 0 \end{pmatrix}, \begin{pmatrix} 0 \\ 0 \end{pmatrix}, \begin{pmatrix} 0 \\ 0 \end{pmatrix} \right)$. Because the number of digits having the same value is by definition k, it must hold for each $d_i : 0 \leqslant d_i^u + d_i^d \leqslant k$.

We distinguish two types of multi-digit error $\mathbf{d}$ corresponding to uncovered (type I) and covered codes (type II) that we can formally describe the following way:

- Type I: $\exists\, i \in \{0, \ldots, i-2\}/d_i^u > d_{i+1}^d$.
- Type II: $\forall\, i \in \{0, \ldots, i-2\} : d_i^u \leq d_{i+1}^d$.

This conjecture can be illustrated by the following two examples:

Example 4 We consider the same hot code 010221 turning into the error code 010220 with the defect $\mathbf{d} = \left(\begin{pmatrix} 0 \\ 0 \end{pmatrix}, \begin{pmatrix} 0 \\ 1 \end{pmatrix}, \begin{pmatrix} 0 \\ 0 \end{pmatrix} \right)$. By comparing the codes before and after the defect happened, we notice that the number of digits '0' has increased, while the number of digits '1' has decreased and the number of digits '2' has remained the same. Any code coinciding with the error-free code at the digits '2' and '1' assigns '1' and '0' to the positions holding '0' in the error code, for instance: 010221 and 110220. All these codes cover the error code because both '1' and '0' cover '0'. This remark can be generalized as follows: whenever the error induces a decrease of the higher-valued digits and an increase in the lower-valued digits, it is always possible to find codes covering the error code. $\square$

Example 5 The opposite case can be illustrated by considering the same code 010221 turning into the error code 010222 with the defect $\mathbf{d} = \left(\begin{pmatrix} 0 \\ 0 \end{pmatrix}, \begin{pmatrix} 1 \\ 0 \end{pmatrix}, \begin{pmatrix} 0 \\ 0 \end{pmatrix} \right)$. Here the number of digits '2' has increased, while the number of digits '1' has decreased and the number of digits '0' has remained the same. Any code that would cover the error code would have at least 3 digits with the value '2'. Such codes do not exist in the considered hot code space. Consequently the error code cannot be covered by any code in the considered space. □

3.3.5.2 Error Type I

The proof for the multi-digit error of type I is given in the following. Consider the case that a code word in Ω is transformed into a code word $\mathbf{c}^*$ by a multi-digit error $\mathbf{d}$ of type I. We denote by i the smallest position in $\mathbf{d}$ at which holds $d_i^u > d_{i+1}^d$. The number of digits in $\mathbf{c}^*$ whose value is $\geq (i+1)$ becomes larger than the permitted number in any code word $\mathbf{c}^b$ in Ω, namely $k \cdot (n - i - 1)$. No code word would imply each digit of $\mathbf{c}^*$ with a value $\geq (i+1)$. Consequently, the nanowire with the pattern equal to the code word $\mathbf{c}^*$ is not addressed by any permitted code word. The probability of defect type I is given by p_I and uses the recursive Algorithm 1:

$$p_I = \sum_{u=0}^{k} \mathsf{ProI}\left(u, 0, 0, \frac{k!}{u! \cdot (k-u)!} \cdot (p_u)^u \cdot (1 - p_u)^{k-u}\right) \qquad (3.2)$$

Here, $u = 0 \ldots k$ is a variable going through all possible numbers of flip-up errors that can happen at digit '0' (flip-downs cannot occur at digit '0'); p_u is the probability of a flip-up error as explained in Sect. 3.3.4; and $\mathsf{ProI}()$ gives the probability that a type I error occurs at any higher-valued digit and that u flip-ups occur at digit '0' (see Algorithm 1).

Algorithm 1 takes as input a defect described by $\begin{pmatrix} x^u \\ x^d \end{pmatrix}$ at the digit level i that can happen with a probability p (describing the history of defects happening at digit levels $\leq i$). It delivers the probability that the assumed defect description can lead to a type I defect. In order to calculate this probability, the algorithm

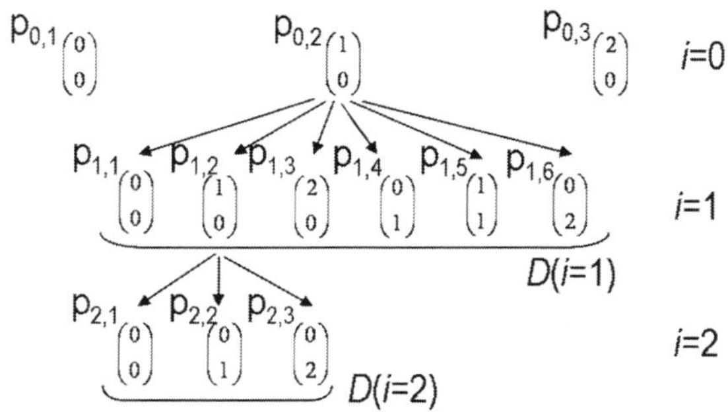

Fig. 3.8 Partial representation of error subtree for error $\begin{pmatrix} 1 \\ 0 \end{pmatrix}$ at $i = 0$

considers all possible error sequences that can happen at digit levels $>i$ and lead to a type I error. The algorithm is explained with the example of Fig. 3.8.

In Fig. 3.8 we assumed a ternary hot code with $M = 6$ and $k = 2$ (e.g. $001122, 012012, \ldots$). Three types of errors can happen at digit '0': $\begin{pmatrix} 0 \\ 0 \end{pmatrix}$, $\begin{pmatrix} 1 \\ 0 \end{pmatrix}$ and $\begin{pmatrix} 2 \\ 0 \end{pmatrix}$, because no flip-downs can happen at this digit. When u goes from 0 to $k = 2$ in Eq. 4.5 it describes these three errors. Each error has the probability $k!/(u! \cdot (k - u)!) \cdot (p_\mathrm{u})^u \cdot (1 - p_\mathrm{u})^{k-u} (u = 0, 1, 2)$, which are denoted by $p_{0,1}, p_{0,2}$ and $p_{0,3}$ respectively in Fig. 3.8. We assume now that Algorithm 4 is called for the error $\begin{pmatrix} 1 \\ 0 \end{pmatrix}$ at the digit '0' (i.e., $u = 1$ in Eq. 4.5). The probability that any possible sequence of errors at higher level digits leads to a type I error is initialized at $\pi_\mathrm{tmp} = 0$ (line 1 in Algorithm 4). We consider the digit level '1', at which 6 different errors can happen: $D(i = 1) = \left\{ \begin{pmatrix} 0 \\ 0 \end{pmatrix}, \begin{pmatrix} 1 \\ 0 \end{pmatrix}, \begin{pmatrix} 2 \\ 0 \end{pmatrix}, \begin{pmatrix} 0 \\ 1 \end{pmatrix}, \begin{pmatrix} 1 \\ 1 \end{pmatrix}, \begin{pmatrix} 0 \\ 2 \end{pmatrix} \right\}$. Generally, for a given level $i \notin \{0, n - 1\}$ the elements of D are generated as follows: $D = \left\{ \begin{pmatrix} x^\mathrm{u} \\ x^\mathrm{d} \end{pmatrix}, \text{s.t.}\ x^\mathrm{d} = 0, \ldots, k \text{ and } x^\mathrm{u} = 0, \ldots, (k - x^\mathrm{d}) \right\}$. For $i = n - 1$, since no flip-ups can happen, $D = \left\{ \begin{pmatrix} 0 \\ x^\mathrm{d}, \end{pmatrix} \text{s.t.}\ x^\mathrm{d} = 0, \ldots, k \right\}$. Every error in $D(i = 1)$ has a probability designated by $p_{1,1} \ldots p_{1,6}$ in Fig. 3.8 and calculated at line 5. We consider the element $\begin{pmatrix} 0 \\ 0 \end{pmatrix}$ in $D(i = 1)$ for which holds: the number of flip-ups at level '0' is higher than the number of flip-downs at level '1'. This error sequence induces a type I error. The algorithm saves the probability of this event (line 9). On the other hand, if we consider the element $\begin{pmatrix} 1 \\ 0 \end{pmatrix}$ in $D(i = 1)$, then no type I error is detected and the algorithm is called iteratively for the next digit level '2' (line 7). The algorithm constructs the elements of $D(i = 2) = \left\{ \begin{pmatrix} 0 \\ 0 \end{pmatrix}, \begin{pmatrix} 0 \\ 1 \end{pmatrix}, \begin{pmatrix} 0 \\ 2 \end{pmatrix} \right\}$ as explained above and goes through them. Only $\begin{pmatrix} 0 \\ 0 \end{pmatrix}$ fulfills the condition of a type I error; for which the error probability is saved (line 16). For the two others 0 is returned (line 14). Then, the updated type I error probability π is returned (line 19 and 21).

Algorithm 1 $p_{\mathrm{I}} = \mathrm{ProI}(x^{\mathrm{u}}, x^{\mathrm{d}}, i, p)$

1: $\pi_{\mathrm{sum}} \leftarrow 0$
2: Construct $D =$ all possible defects in subtree
3: **for all** $\mathbf{y} = (y^{\mathrm{u}}, y^{\mathrm{d}}) \in D$ **do**
4:     **if** $i < n - 1$ **then**
5:         $p \leftarrow p \cdot k!/(y^{\mathrm{d}}! \cdot y^{\mathrm{u}}! \cdot (k - y^{\mathrm{d}} - y^{\mathrm{u}})!) \cdot (p_{\mathrm{u}})^{y^{\mathrm{u}}} \cdot (p_{\mathrm{d}})^{y^{\mathrm{d}}} \cdot (1 - p_{\mathrm{u}} - p_{\mathrm{d}})^{k - y^{\mathrm{u}} - y^{\mathrm{d}}}$
6:         **if** $y^{\mathrm{d}} \geq x^{\mathrm{u}}$ **then**
7:             $\pi \leftarrow \mathrm{ProI}(y^{\mathrm{u}}, y^{\mathrm{d}}, i + 1, p)$
8:         **else**
9:             $\pi \leftarrow p$
10:        **end if**
11:    **else**
12:        $p \leftarrow p \cdot k!/(y^{\mathrm{d}}! \cdot (k - y^{\mathrm{d}})!) \cdot (p_{\mathrm{d}})^{y^{\mathrm{d}}} \cdot (1 - p_{\mathrm{d}})^{k - y^{\mathrm{d}}}$
13:        **if** $y^{\mathrm{d}} \geq x^{\mathrm{u}}$ **then**
14:            $\pi \leftarrow 0$
15:        **else**
16:            $\pi \leftarrow p$
17:        **end if**
18:    **end if**
19:    $\pi_{\mathrm{sum}} \leftarrow \pi_{\mathrm{sum}} + \pi$
20: **end for**
21: **return** π_{sum}

3.3.5.3 Error Type II

Now, let the code word $\mathbf{c}^a$ in Ω be transformed into $\mathbf{c}^*$ by a multi-digit error $\mathbf{d}$ of type II. The number of digits in $\mathbf{c}^*$ having the value i is always larger than the number of digits having the value $(i + 1)$. It is possible to construct one or more code words $\mathbf{c}^b$ in Ω implying $\mathbf{c}^*$. An intuitive way consists in starting with the smallest digit value 0, and filling the digits of $\mathbf{c}^b$ by 0 with respect to the positions hold by the value 0 in the digits of $\mathbf{c}^*$. The procedure is repeated iteratively on next digit values until all digits of $\mathbf{c}^b$ are allocated (Algorithm 2). In Algorithm 2 we use the following notations: the number of digits having the value i in $\mathbf{c}^*$ is l_i. Their respective positions are $p_0^i, \ldots, p_{l_i-1}^i$. The definition of the defect pattern $\mathbf{d}$ yields: $l_i = k - d_i^u + d_{i+1}^d + d_{i-1}^u \, \forall i$, with the convention $d_{i-1}^u = 0$ for $i = 0$ and $d_{i+1}^d = 0$ for $i = n - 1$.

From the definition of the type II errors, there is at least a digit value i for which $v_i < l_i$ holds. Each choice of v_i elements among l_i possible values (line 2) gives many possible choices for $\mathbf{c}^b$. This proves that it is possible to find more than one code word covering the considered error code type II. The following example explains Algorithm 2 by means of the error code $\mathbf{c}^* = 010220$:

Example 6 In order to find all code words which cover a given error code $\mathbf{c}^*$, Algorithm 2 is called with $i = 0, v = k, \Delta$ a single space-holder code with $M \times$ '*', where '*' means that the digits is not allocated yet and can take any possible value. For $\mathbf{c}^* = 010220$, Algorithm 2 is called with $i = 0, v = 2$ and $\Delta = \{******\}$. We start with the digit level '0', whose positions in $\mathbf{c}^*$ are given by $S_1 = \{0, 2, 5\}$ (line 1). Then, $\sigma = \{\{0, 2\}, \{0, 5\}, \{2, 5\}\}$ (line 2). Lines 4–12 set Δ_{tmp} to $\{0 * 0 * *1, 0 * 1 * *0, 1 * 0 * *0\}$. Line 16 calls the algorithm with $i = 1, v = 1$ and Δ set to Δ_{tmp}. During this call of Algorithm 2, $S_1 = \{1\}$ (line 1) and S_2 can take the only value $\{1\}$. Then, lines 4–12 update Δ_{tmp} to $\{010 * *1, 011 * *0, 110 * *0\}$. Subsequently, the algorithm is called again with $i = 2, v = 2$ and Δ set to Δ_{tmp} (line 16). During this call, the lines 17–20 are executed and just fill in the remaining space-holders with the digit '2'. Finally, Δ is set to $\{010221, 011220, 110220\}$ and returned recursively to the top level. To conclude, the error code $\mathbf{c}^*$ can be covered by 3 possible code words. □

Algorithm 2 $\Sigma = \texttt{CoveredSet}(i, \nu, \Delta, \mathbf{c}^*)$

1: Construct $S_1 = $ set of all positions of digit i in $\mathbf{c}^*$
2: Construct $\sigma = $ set of subsets of S_1 with ν elements
3: **if** $i < n - 1$ **then**
4:    $\Delta_{\text{tmp}} = \varnothing$
5:    **for all** $S_2 \in \sigma$ **do**
6:       $\Delta_{\text{rep}} = \Delta$
7:       **for all** $\mathbf{c}^b \in \Delta_{\text{rep}}$ **do**
8:          Allocate i to digits of $\mathbf{c}^b$ at positions S_2
9:          Allocate $i + 1$ to digits of $\mathbf{c}^b$ at positions $S_1 \backslash S_2$
10:       **end for**
11:       $\Delta_{\text{tmp}} \leftarrow \Delta_{\text{tmp}} \cup \Delta_{\text{rep}}$
12:    **end for**
13:    $\nu \leftarrow \nu + k - |S_1|$
14:    $i \leftarrow i + 1$
15:    $\Delta \leftarrow \Delta_{\text{tmp}}$
16:    **return** $\texttt{CoveredSet}(i, \nu, \Delta, \mathbf{c}^*)$
17: **else**
18:    **for all** $\mathbf{c}^b \in \Delta$ **do**
19:       Allocate $n - 1$ to digits of $\mathbf{c}^b$ at remaining positions
20:    **end for**
21:    **return** Δ
22: **end if**

3.3.5.4 Immune Code Space

For simulation purposes that will be explained in Sect. 3.3.8, it is useful to know the size of the code space that did not undergo any defects, i.e., the probability that a code word is immune: p_{im}. In the **k**-hot code space with the length M and $\mathbf{k} = (k, \ldots, k)$, there are k digits having the value 0, which can undergo only flip-up defects; thus, the probability of each one of them of being error-free is $1 - p_u$. On the other hand, there are k digits having the value $n - 1$, which can undergo only flip-down defects; thus, the probability of each one of them of being error-free is $1 - p_d$. The remaining $(n - 2)k$ digits can undergo both flip-up and flip-down defects and the probability of each one of them of being error-free is: $1 - p_u - p_d$. Consequently, the probability that the whole hot code word is error-free is:

$$p_{im} = (1 - p_u - p_d)^{(n-2)k} \cdot (1 - p_u)^k \cdot (1 - p_d)^k$$

3.3.5.5 Unique Covering

In order to assess the probability that a code word $\mathbf{c}^a$ uniquely covers an error code, we first need to enumerate the code words $\mathbf{c}^b$ which can undergo a sequence of defects to become covered by $\mathbf{c}^a$ and the probability of each one of these events. Then, we can derive the probability that exactly one of these events happens; which is equivalent to saying that $\mathbf{c}^a$ covers a *unique* error code.

Enumerating these events is performed in two steps: We first define the set S of code words $\mathbf{c}^b$ that can be transformed into $\mathbf{c}^a$ by a sequence of flip-ups and -downs. Then, the considered events consist in making each element $\mathbf{c}^b$ undergo a sequence of defects that turn it not necessarily into $\mathbf{c}^a$ but just make it covered by $\mathbf{c}^a$. We describe the elements $\mathbf{c}^b$ of S in an abstract way consisting in a transformation matrix $\mathcal{T}$ which describes the flip-ups and -downs that $\mathbf{c}^b$ needs to undergo in order to turn to $\mathbf{c}^a$. Then, each element is assigned the probability of covering under defects.

A code word $\mathbf{c}^b$ is transformed into another code word $\mathbf{c}^a$ by undergoing at each digit level $i = 0 \ldots n - 1$: t_j^i $(j - i)$-order flip-ups (for $j = i + 1 \ldots n - 1$) and t_j^i $(i - j)$-order flip-downs (for $j = 0 \ldots i - 1$). We use the following convention: t_i^i designates the number of correct digits at the level i. A transformation $\mathcal{T}$ affecting a whole code word is a set of the transformations $\mathbf{t}^i = [t_0^i, \ldots, t_{n-1}^i]^\top$ affecting each digit level $i = 0 \ldots n - 1$. Thus, $\mathcal{T}$ is the matrix $[\mathbf{t}^0, \ldots \mathbf{t}^{n-1}]$. The $(i + 1)$-th column of $\mathcal{T}$ describes the transformation happening at digits level $i = 0 \ldots n - 1$. Since there are k occurrences of the digit level i in each code word, $\mathcal{T}$ must verify: $\forall i \sum_j t_j^i = k$. The $(i + 1)$-th row indicates the number of digits with the value i in the code obtained after the transformation (i.e., $\mathbf{c}^a$). This number has to be set to k. Then, we derive the second condition on $\mathcal{T}$: $\forall j \sum_i t_j^i = k$.

Example 7 We consider the ternary hot code with $M = 6$ and $k = 2$. The digit level '0' undergoes only one flip-up: $\mathbf{t}^0 = [1, 1, 0]^\top$. The digit level '1' undergoes one flip-up and one flip-down: $\mathbf{t}^1 = [1, 0, 1]^\top$. The digit level '2' undergoes no errors: $\mathbf{t}^2 = [0, 0, 2]^\top$. Thus the code word 001122 undergoing $\mathcal{T} = [\mathbf{t}^0, \mathbf{t}^1, \mathbf{t}^2]$, can be transformed, for instance into 010222; which is not an element of the considered code space. This is due to the fact that the second condition is not fulfilled. We suggest the transformation $\widetilde{\mathcal{T}}$ meeting both conditions:

$$\mathcal{T} = \begin{bmatrix} 1 & 1 & 0 \\ 1 & 0 & 0 \\ 0 & 1 & 2 \end{bmatrix} \quad \widetilde{\mathcal{T}} = \begin{bmatrix} 1 & 1 & 0 \\ 1 & 0 & 1 \\ 0 & 1 & 1 \end{bmatrix}$$

$\widetilde{\mathcal{T}}$ would transform 001122 into 010212 (for instance), which is in the same code space.   □

Now, we inject defects that make $\mathbf{c}^b$ controlled by $\mathbf{c}^a$. At each digit value i, the t^i_j digits, for which $j > i$ holds, must undergo a $(j - i)$-order (or higher) flip-down defect each, in order to decrease their value down to that of the corresponding digit in $\mathbf{c}^a$. On the contrary, the t^i_j digits, for which $j < i$ holds, already have a value which is smaller than that of the corresponding digit in $\mathbf{c}^a$, so they can undergo any kind of first-order defect or be correct. The remaining t^i_i digits in $\mathbf{c}^b$ have the same values as their counterparts in $\mathbf{c}^a$, so they can undergo a flip-down error or no error. Only first order flip errors are considered, because the likelihood of higher order defects is negligible. Thus, we set t^i_j to 0 for $j > i + 1$ in order to reduce the computation time. This represents the third condition on $\mathcal{T}$.

These conditions define the eligible set of transformations $\mathcal{T}$. For each transformation, the probability that a sequence of defects transforms the digits of $\mathbf{c}^b$ with the value i into an error code covered by $\mathbf{c}^a$ at the positions corresponding to these digits is: $\tilde{p}_i = (p_d)^{u_i} \cdot (1 - p_u)^{k - u_i - d_i}$ with $u_i = t^i_{i+1}$ and $d_i = \sum_{j < i} t^i_j \forall i$. The probability that the code word $\mathbf{c}^b$ is transformed into an error code covered by $\mathbf{c}^a$ is identical to the probability that the previous event holds for each digit value i. This gives the event probability $p = \prod_i \tilde{p}_i$.

Given the transformation $\mathbf{t}^i$ affecting the k occurrences of the digit value i, there are μ_i distributions of the t^i_j transformations on the k elements, given by: $\tilde{\mu}_i = k! / \prod_j t^i_j!$. Considering the n possible values of i, there are μ code words $\mathbf{c}^b$ transformed into an error code covered by $\mathbf{c}^a$ with the same probability $p : \mu = \prod_i \tilde{\mu}_i$.

Example 8 The code $\mathbf{c}^a = 001122$ in the previous example was transformed by $\widetilde{\mathcal{T}}$ to $\mathbf{c}^b = 010212$. A series of defects transforms $\mathbf{c}^b$ into $\mathbf{c}^*$ that is covered by $\mathbf{c}^a$. Thus the 1st digit in $\mathbf{c}^b$ (corresponding to the 1st occurrence of '0' in $\mathbf{c}^a$) has to remain unchanged in order to be covered by $\mathbf{c}^a$. The 2nd digit in $\mathbf{c}^b$ (corresponding to the 2nd occurrence of '0' in $\mathbf{c}^a$) has to flip down to be covered by $\mathbf{c}^a$. This gives

$\tilde{p}_0 = p_{\mathrm{d}}(1 - p_{\mathrm{u}})$. In a similar way, we find $\tilde{p}_1 = p_{\mathrm{d}}$ and $\tilde{p}_2 = 1$; Then $p = (p_{\mathrm{d}})^2(1 - p_{\mathrm{u}})$. The are many images of $\mathbf{c}^{\mathrm{a}}$ through $\widetilde{\mathcal{T}}$; for instance 102012, 102021 etc. Their number is obtained by permuting the flipping digits. For instance $\widetilde{\mathcal{T}}$ requires that exactly one digit '0' flips up: there are $\widetilde{\mu}_0 = 2!/(1! \cdot 1! \cdot 0!) = 2$ permutations. In a similar way, we find $\widetilde{\mu}_1 = 2$ and $\widetilde{\mu}_2 = 2$ for the digits '1' and '2' respectively. Finally, the number of possible code words $\mathbf{c}^{\mathrm{b}}$ is $\mu = 8$. ☐

Algorithm 3 represents a formulation of the method explained above. It assumes that all eligible transformations $\mathcal{T}$ were calculated, for instance, by selecting in an exhaustive way those meeting the 3 conditions mentioned above among all matrices in $0, \ldots, k^{n \times n}$. Then, it defines the set S of all (p, μ); where p is the probabilities of the code words $\mathbf{c}^{\mathrm{b}}$ represented by the valid transformation to undergo the right error sequence that makes it covered by $\mathbf{c}^{\mathrm{a}}$, and μ is the number of its equivalent occurrences.

Let S be the set of all eligible events with their respective occurrences; then, the probability that exactly one event happens (i.e., $\mathbf{c}^{\mathrm{a}}$ covers exactly one error code) can be calculated as follows:

$$p_{\mathrm{U}} = \sum_{i=1,\ldots,|S|} \mu_i \cdot p_i/(1 - p_i) \times \prod_{i=1,\ldots,|S|} (1 - p_i)^{\mu_i} \qquad (3.3)$$

Algorithm 3 $S = \mathtt{UniqueSet}(\mathcal{T})$

1: Construct $\mathcal{E} = \{$all eligible transformations $\mathcal{T}$ meeting the 3 conditions above$\}$
2: $S \leftarrow \varnothing$
3: **for all** $\mathcal{T} \in \mathcal{E}$ **do**
4:     $u_i \leftarrow t^i_{i+1} \, \forall \, i$
5:     $d_i \leftarrow \sum_{j<i} t^i_j \, \forall \, i$
6:     $\tilde{p}_i \leftarrow p_{\mathrm{d}}^{u_i} \cdot (1 - p_{\mathrm{u}})^{k - u_i - d_i}$
7:     $p \leftarrow \prod_i \tilde{p}_i$
8:     $\tilde{\mu}_i \leftarrow k!/\prod_j t^i_j!$
9:     $\mu \leftarrow \prod_i \tilde{\mu}_i$
10:    $S \leftarrow S \cup \{(p, \mu)\}$
11: **end for**
12: **return** S

In summary, this section suggested mathematical methods to estimate the probability of uncovered codes (p_{I}) and the probability of unique covering (p_{U}). The probability p_{I} was calculated by constructing the whole set of type I errors and estimating the probability of each one of these errors. On the other hand, p_{U} was estimated from the set of code words that can undergo a sequence of errors and become covered by a given reference code word. Once p_{I} and p_{U} are known, they can be inserted in Eq. 4.4 in order to estimate the size of the addressable code space under defects, i.e., the number of nanowires that can be addressed in the array under high variability conditions.

3.3.6 Errors in the NRC Space

3.3.6.1 Error Types

In the following Ω refers to an arbitrary NRC space with length M (M is even) and base size n. We define a flip-up defect at digit c_i in the code word $\mathbf{c}$ to be canceled when a flip-down defect occurs at digit $c_{M/2+i}$. A canceled flip-down defect is defined in a complementary way.

As for the hot codes, we distinguish two types of multi-digit errors for reflexive codes, corresponding to the uncovered (type I) and covered codes (type II) that we can describe in the following way:

- Multi-digit errors of type I: The code word undergoes at least one uncanceled flip-up.
- Multi-digit errors of type II: The code word only experiments flip-downs and/or canceled flip-ups.

In order to illustrate this assumption, we suggest the following two examples:

Example 9 We consider the code word $\mathbf{c}^a = 00012221$ in the ternary ($n = 3$) reflexive code space with the length $M = 8$. $\mathbf{c}^a$ undergoes an uncanceled flip-up error at the fourth digit, and turns to the error code $\mathbf{c}^* = 00022221$. Any hypothetical code word that would cover $\mathbf{c}^*$ would have at the forth digit the value 2 and at the last digit the value 0 (then the last digit would not covered); or it would have at the last digit the value 1 or 2 and at the fourth digit the value 1 or 0 respectively (then, in both cases, the fourth digit would not be covered). The uncanceled flip-up defect causes any hypothetical covering code to cover either the first half of the code or its reflected half, but never both of them at the same time. Consequently, the error code $\mathbf{c}^*$ cannot be covered by any code in the considered NRC space.   □

Example 10 On the other hand, if we consider the same code word $\mathbf{c}^a$ undergoing the canceled flip-up defect at the fourth digit, then it turns to $\mathbf{c}^* = 00022220$ which is a code word from the same space and it is consequently covered by itself. If, in addition to the canceled flip-up defect at the fourth digit a flip-down defect occurs, say at the fifth digit, then the code turns to the error code $\mathbf{c}^* = 00021220$ which is in turn covered by the code words 00022220 and 10021220.   □

3.3.6.2 Error Type I

Here, we explain formally how codes undergoing type I errors are not covered by any code in the considered code space. Let the code word $\mathbf{c}^a$ in Ω be transformed into $\mathbf{c}^*$ by a multi-digit error of type I. We denote by i one of the positions in the first half of the code word, at which an uncanceled flip-up error occurs. Then, $c_i^* = c_i^a + 1$ and $c_{M/2+i}^* \geq c_{M/2+i}^a$. Any code word $\mathbf{c}^0$ which would imply the pattern corresponding to $\mathbf{c}^*$ would verify $c_i^0 \geq c_i^* = c_i^a + 1$. Then, $c_{M/2+i}^0 = \mathsf{NOT}(c_i^0) = n - 1 - c_i^0 \leq n - 1 - c_i^a - 1 = c_{M/2+i} - 1 < c_{M/2+i}^*$ and $\|\sigma(\mathbf{c}^* - \mathbf{c}^0)\| \neq 0$. Thus, there exists no code word in Ω that would cover $\mathbf{c}^*$. The size of the addressable space is reduced by the number of code words undergoing the multi-digit error of type I.

Unlike hot codes, we give no recursive form for the probability of type I errors affecting reflexive codes, we rather derive the explicit analytical expression. However, it is much easier to consider the complementary case of the type I error, which is the type II error (probability p_{II}) or no errors (immune codes with the probability p_{im}): $p_I = 1 - (p_{II} + p_{im})$. The exact expression of $p_{II} + p_{im}$ and consequently the one of p_I are derived in the following subsection.

3.3.6.3 Error Type II

Let the code word $\mathbf{c}^a$ in Ω be transformed into $\mathbf{c}^*$ by a multi-digit error $\mathbf{d}$ of type II. It is possible to find more than one code word $\mathbf{c}^0$ which covers the pattern corresponding to $\mathbf{c}^*$. We construct first, say, the left half of $\mathbf{c}^0$, then the right half is obtained by complementing the left half. The construction rule is the following $(i = 0, \ldots, M/2 - 1)$:

- If c_i^a undergoes a non-canceled flip-down error, then c_i^0 can be set to either $c_i^a - 1$ or c_i^a.
- If c_i^a undergoes a canceled flip-down error, then c_i^0 is set to $c_i^a - 1$.
- If c_i^a undergoes a canceled flip-up error, then c_i^0 is set to $c_i^a + 1$.
- If c_i^a has no error, then c_i^0 is set to c_i^a.
- $c_{M/2+i}^0$ is set to $n - 1 - c_i^0$

It is easy to verify that all patterns corresponding to $\mathbf{c}^0$ constructed this way cover $\mathbf{c}^*$. In order to calculate the probability p_I of the multi-digit error type I, we observe the complementary event (multi-digit error type II or no error in code word $\mathbf{c}^a$):

- If c_i^a undergoes no error, then $c_{M/2+i}^a$ must undergo a flip-down or no error: $(1 - p_u - p_d)(1 - p_u)$.
- If c_i^a undergoes a flip-down error, then $c_{M/2+i}^a$ can have any value: p_d.
- If c_i^a undergoes a flip-up error, then $c_{M/2+i}^a$ must undergo a flip-down error: $p_u p_d$.

This scheme assumes that the digit c_i^a have the value $1 \ldots n - 2$, i.e., it is not at the range borders. In this case, the probability that the digit c_i^a and its complement in the reflected code half undergo no type I error is:

$$\overline{p}_{I,1} = (1 - p_u - p_d)(1 - p_u) + p_d + p_u p_d$$

If the considered digit c_i^a has the value $n - 1$, then its complement $c_{M/2+i}^a$ has the value 0, and the conditions above become:

- If c_i^a undergoes no error, then $c_{M/2+i}^a$ cannot undergo a flip-down error, it must be correct: $(1 - p_d)(1 - p_u)$.
- If c_i^a undergoes a flip-down error, then $c_{M/2+i}^a$ can have any value: p_d.
- c_i^a cannot undergo a flip-up error, then this case has the probability 0.

Thus, the probability that the digit $n - 1$ and its complement in the reflected code half undergo no type I error is given by:

$$\overline{p}_{I,2} = (1 - p_d)(1 - p_u) + p_d$$

In a similar way, we find that the probability that the digit with the value 0 and its complement in the reflected code half undergo no type I error is:

$$\overline{p}_{I,3} = 1 - p_u + p_u p_d$$

We notice that $\overline{p}_{I,2} = \overline{p}_{I,3}$, which is due to the reflexive principle. By averaging these probability (application of the binomial form), we can estimate the probability that any digit and its complement in the reflected code half undergo no type I error:

$$\overline{p}_{I,avg} = \overline{p}_{I,1} \cdot (n - 2)/n + \overline{p}_{I,2}/n + \overline{p}_{I,3}/n$$

Since a code word has $M/2$ couples of complementary digits, the probability that no type I error happens, i.e., that no error or only type II errors happen is given by:

$$p_{II} + p_{im} = \overline{p}_{I,avg}^{M/2}$$

Then, the value of p_I can be given as $1 - (p_{II} + p_{im})$:

$$p_I = 1 - \left(\frac{2}{n}(1 - p_u + p_u p_d) + \frac{n - 2}{n}(p_u^2 - 1) \right)^{M/2} \tag{3.4}$$

3.3.6.4 Immune Code Space

Now, we consider the part of the code space that did not undergo any defects in order to calculate the probability of an immune code word p_{im}. The number of pairs of digits at the range borders can be any natural number between 0 and $M/2$. Because of the reflection principle, every pair among them has one digit with the value 0 and one digit with the value $n - 1$. Then, the probability of this digit pair to be error-free is $(1 - p_u)(1 - p_d)$. The error-free probability of each one of the remaining $M/2 - i$ digit pairs, which can undergo either flip-up or flip-down errors, is given by $(1 - p_u - p_d)^2$. Finally the probability that the whole code word is error-free is given by:

$$p_{im} = \sum_{i=0}^{M/2} \binom{M/2}{i} \left(\frac{2}{n}\right)^i \left(\frac{n-2}{n}\right)^{M/2-i}$$
$$\times \left((1 - p_u - p_d)^2\right)^{M/2-i} ((1 - p_u)(1 - p_d))^i$$

By using the binomial form, this expression can be simplified to the weighted sum of the case that the pair is at the range border and the case that it is not:

$$p_{im} = \left(\frac{2}{n}(1 - p_u)(1 - p_d) + \frac{n-2}{n}(1 - p_u - p_d)^2\right)^{M/2}$$

3.3.6.5   Unique Covering

Now, we would like to calculate p_U. In principle, the same reasoning for hot codes can be applied on reflexive codes too. In order to assess the probability that a code word $\mathbf{c}^a$ uniquely covers an error code, we first enumerate all the code words $\mathbf{c}^b$ that can be transformed into $\mathbf{c}^a$ by a sequence of flip-ups and -downs. Then, we calculate the probability that these code words $\mathbf{c}^b$ undergo errors and become covered by $\mathbf{c}^a$.

For hot codes, we considered all the transformations $\mathcal{T} = (\mathbf{t}^0, \ldots, \mathbf{t}^{n-1})$ that transform $\mathbf{c}^b$ into $\mathbf{c}^a$ and we represented them by matrices. The fact that we considered only first-order defects fixed the limit $t_j^i = 0$ for $j > i + 1$. Because of the reflection principle, the reflected half of the code word fixes the limit $t_j^i = 0$ for $j < i - 1$. consequently the transformation becomes much simpler than in the case of hot codes: for each digit i, $(c_i^a - c_i^b) \in \{-1, 0, 1\}$ must hold. We do not need the matrix representation in order to define the set of code words $\mathbf{c}^b$ verifying this condition: indeed, the set of $\mathbf{c}^b$ represents the neighborhood of $\mathbf{c}^a$ that can be graphically represented in the space $\mathbb{B}^{M/2}$ by a hypercube with the edge length 2 and centered around $\mathbf{c}^a$ (see Fig. 3.9). If $\mathbf{c}^a$ has a certain number u of digits at the range border (i.e., 0 or $n - 1$), then only $(c_i^a - c_i^b) \in \{-1, 0\}$ or $\{0, 1\}$ holds respectively; and the volume of hypercube is halved for each one of these digits.

Given this definition of the set of code words $\mathbf{c}^b$, we can now enumerate them, first, by assuming that no digit is at the range borders. We consider only the first half of the code words because it completely defines the whole code word by

Fig. 3.9 Hypercube representing the set of code words $\mathbf{c}^b$ that can be covered by $\mathbf{c}^a$ when they undergo defects

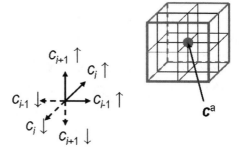

applying the reflection principle. Let α be the number of digits in a half-word such that $c_i^b - c_i^a = 0(\alpha = 0, \ldots, M/2)$. There are $\tilde{\mu}_\alpha = \begin{pmatrix} M/2 \\ \alpha \end{pmatrix} \cdot 2^{M/2-\alpha}$ possible code words fulfilling this condition. Each one of them can be transformed into an error code $\mathbf{c}^*$ covered by $\mathbf{c}^a$ if the α digits in each half undergo no flip-up error, and each one of the $M/2 - \alpha$ digits for which holds $c_i^* - c_i^a = 1$ (in the whole code word) undergoes a flip-down error. The other $M/2 - \alpha$ digits for which holds $c_i^* - c_i^a = -1$ (in the whole code word) can undergo any kind of defect or they can be correct. The likelihood of each event is $\tilde{p}_\alpha = (1 - p_u)^{2\alpha} \cdot p_d^{M/2-\alpha} \cdot 1^{M/2-\alpha}$.

We can now consider the cases in which the first half of $\mathbf{c}^b$ has u digits at the range border such that $c_i^b - c_i^a \neq 0$ and α is still the number of digits in the half-word such that $c_i^b - c_i^a = 0(\alpha = 0, \ldots, M/2$ and $u = 0, \ldots, M/2 - \alpha)$. The number of code words having u digits among $M/2 - \alpha$ such that $c_i^b - c_i^a \neq 0$ is $2^{(M/2-\alpha)-u} \cdot 1^u$. The average number of occurrences of each one of these words is $\begin{pmatrix} M/2 - \alpha \\ u \end{pmatrix} (2/n)^u ((n-2)/n)^{M/2-\alpha-u}$. Then, the average number of code words having α digits such that $c_i^b - c_i^a = 0$ is:

$$\mu_\alpha = \begin{pmatrix} M/2 \\ \alpha \end{pmatrix} \sum_{u=0}^{M/2-\alpha} \left(2^{M/2-\alpha-u} \times \begin{pmatrix} M/2 - \alpha \\ u \end{pmatrix} \left(\frac{2}{n}\right)^u \left(\frac{n-2}{n}\right)^{M/2-\alpha-u} \right)$$

This expression can be explicitly calculated by using the binomial formula, yielding the following:

$$\mu_\alpha = \begin{pmatrix} M/2 \\ \alpha \end{pmatrix} \left(2 \cdot \frac{n-1}{n} \right)^{M/2-\alpha}$$

We notice that the impact of the digits at the range border on the number of code words corresponding to a given α is the additional factor $(n-1)/n$.

We consider now the probability of occurrence of the transformation affecting each one of the μ_α code words without discarding the digits at the range borders. The indexes α and u keep the same definitions as before (α being again the number of digits in the half of a code word such that $c_i^b - c_i^a = 0$, and u the number of digits in the rest of the code half having their values at the range borders: $\alpha = 0, \ldots, M/2$ and $u = 0, \ldots, M/2 - \alpha$). In addition we call v the number of digits in the set of μ digits having their value at the range borders ($v = 0, \ldots, \alpha$). First, we fix the value of α. In order for $\mathbf{c}^b$ to undergo errors making it covered by $\mathbf{c}^a$, each one of the α digits must undergo no flip-up error. If the digit is among the ones at the range border (with the probability $2/n$), then having no flip-up error happens with the probability $1^v \cdot (1 - p_u)^v$. If the digit is not at the range border (with the probability $(2 - n)/n$), then having no flip-up error happens with the probability $(1 - p_u)^{2(\alpha-v)}$. The remaining $M/2 - \alpha$ digits are divided into those for which holds $c_i^b - c_i^a = 1$, and which have to undergo a flip-down error, and those for which holds $c_i^b - c_i^a = -1$, and which can undergo any kind of error or be correct. Because of the reflection principle, the probability that the remaining $M/2 - \alpha$ digits undergo the right error sequence is independent of u and it is equal to $p_d^{M/2-\alpha}$. Consequently, the probability that each one of the μ_α code words $\mathbf{c}^b$ undergoes the right defect sequence and becomes covered by $\mathbf{c}^a$ is given by:

$$p_\alpha = \sum_{v=0}^{\alpha} \binom{\alpha}{v} \left(\frac{2}{n}\right)^v \left(\frac{n-2}{n}\right)^{\alpha-v} \times (1 - p_u)^v (1 - p_u)^{2(\alpha-v)} p_d^{M/2-\alpha}$$

This expression can be explicitly calculated by using the binomial form and gives:

$$p_\alpha = \left(\frac{n-2}{n}(1 - p_u)^2 + \frac{2}{n}(1 - p_u)\right)^\alpha \cdot p_d^{M/2-\alpha}$$

We notice that the value of p_α is just the weighted average of the cases whether the digit is at the range borders or not.

Finally, having the set S of all possible events that $\mathbf{c}^b$ becomes covered by $\mathbf{c}^a$ and their respective occurrences, the likelihood that only one event happens, i.e., only one code word in S is transformed into another word covered by $\mathbf{c}^0$ is given by:

$$p_U = \sum_{i=0}^{M/2} \mu_i \cdot p_i / (1 - p_i) \times \prod_{i=0}^{M/2} (1 - p_i)^{\mu_i} \tag{3.5}$$

In summary, this section suggested the analytical expressions of the probability of uncovered codes (p_I) and the probability of unique covering (p_U). For each probability, we considered the simplified case, in which the digits are not at the range border, i.e., 0 and $n - 1$, then we derived the general case, where the digits can have any value. The NRC space is easier to describe analytically; thus, the given probabilities are in explicit forms unlike the recursive forms given in the k-hot code space. Once p_I and p_U are known, they can be used in Eq. 4.4 in order to analytically estimate the size of the addressable code space under defects conditions.

3.3.7 Assumptions of the Simulations

The defect models defined in Sect. 3.3.4 enable the assessment of the number of nanowires that the decoder can address uniquely even with variable threshold voltages. Based on this estimate, we can evaluate the effective capacity of the crossbar circuit, i.e., the number of addressable crosspoints. In order to keep the circuit used for simulations simple, we assumed that the crossbar operates as a memory. The memory effective capacity, i.e., the number of addressable crosspoints, is used as a metrics to evaluate the decoder and its impact on yield. Consequently, defects affecting the molecular switches were not included in the model. Then, we can estimate the effective capacity C_{eff} in a similar way to the memory yield in [1]: $C_{eff} = (\eta P_{contact}^2 \cdot |\Omega'|)^2$, with $P_{contact}$ the probability that the nanowire ohmic contact is good ($P_{contact} = 0.95$ from [1]). In this expression, η is a statistic parameter that depends on the nanowire fabrication process. The area can be estimated from the geometry of the layout. Furthermore, we used the following simulation parameters: $q = 100, V_0 = V_{DD} = 1\,\text{V}, \alpha \sim 1/n$ since the difference between the highest and lowest $V_{T,i}$'s is limited by the constant V_{DD}. The following sections explain how the area and the value of η are derived.

3.3.7.1 Assumptions on the Circuits Geometry

The circuit geometry is assumed to be the same as the one explained in Sect. 3.1. The mesowires are defined according to the technology node 45 nm (DRAM half-pitch as estimated for 2010 in the ITRS review of 2007 [29]). We would highlight the assumptions resulting from the fact that we only consider feasible designs from the point of view of the physical designer (see Fig. 3.10):

- The polycrystalline silicon half pitch is designated by f.
- The contact width ($W_{contact}$) is equal to the overlap area between the poly-Si and metal 1 layers. Although it is difficult to predict the contact width for highly scaled technologies depending on f in a generic way without any reference to a specific fabrication plant, we estimated it to be $2 \times f$.
- The mesowire pitch is equal to the sum of the contact width and the poly-Si half-pitch, i.e., $3 \times f$.
- The width of a contact group is the sum of the contact width and $2\times$ the poly-silicon half-pitch; i.e., $4 \times f$.

The nanowire thickness and pitch are taken from the references in which the considered decoders were presented and we used the best proven values:

- Grown nanowires: the nanowire core has a thickness of 5 nm, An additional insulator shell adds 4 nm to the nanowire thickness, making the pitch = 9 nm [2].

Fig. 3.10 Design-rule-
accurate layout of crossbar
memories used for yield
estimation of bottom-up and
top-down approaches

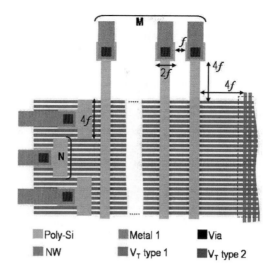

- Nanowires patterned with nano-imprint: the nanowire thickness is around 6 nm, and the pitch is around 13 nm [10].

3.3.7.2 Statistical Assumptions

Nanowires grown in a bottom-up process have several sources of defects and uncertainty due to their possible sublithographic pitch. These defects were analyzed in [1]. The impact of all these additional defects is modeled as a factor η and the effective memory capacity for bottom-up approaches becomes:

$$C_{\text{eff}}^{\text{BU}} = \left(\eta^{\text{BU}} \cdot P_{\text{contact}}^2 \cdot |\Omega'| \right)^2$$

We used the same sources of defects as those modeled in [1] and [2]:

- Probability of non-broken nanowires. P_{nbr} is dependant on the nanowire length, and for 10 μm long nanowires, $P_{\text{nbr}} = 0.90$.
- Probability of unique nanowires. P_{unq} is the probability that each nanowire in a contact group is unique. It depends on the code space size and the way it can be deduced is explained in [1].
- Probability of a good control of nanowires P_{cnt}. The nanowires may be displaced with respect to each other after the fluidic deposition process. The average value 0.80 was used as suggested in [2].
- Probability of no nanowire loss at the interfaces between contact groups. P_{int} was estimated to be $(1 - 1/N)$, where N is the number of nanowires in each contact group.

This factor η^{BU} represents the impact of all these defects, thus it was set to the product of their probabilities:

$$\eta^{BU} = P_{nbr} \cdot P_{unq} \cdot P_{cnt} \cdot P_{int}$$

On the contrary, nanowires produced in a top-down approach rarely suffer from breakage $(P_{nbr} = 1)$. Their codes are not assigned randomly, but by using lithography $(P_{unq} = 1)$. Their ohmic contact was not reported to cause electrical loss since the nanowires are not dispersed randomly on the substrate to be functionalized $(P_{cnt} = 1)$. For nanowires defined on the lithographic pitch, i.e., those fabricated with standard lithography, there is no interface loss $(P_{int} = 1)$. However, those defined below the photolithography limit, for instance by using nanomolds, may have an interface loss quantized as suggested previously: $P_{int} = 1 - 1/N$, with N the number of nanowires to be addressed in the contact region. Then, for top-down nanowires we have:

$$C_{eff}^{TD} = \left(\eta^{TD} \cdot P_{contact}^2 \cdot |\Omega'| \right)^2$$

with $\eta^{TD} = 1$ for top-down nanowires defined on the lithography pitch, and $\eta^{TD} = 1 - 1/N$ for top-down nanowires defined on a sub-lithographic pitch.

3.3.8 Simulations of the Addressable Code Space

In order to assess the variation of the addressable code space under variable V_T, we plotted separately the uncovered part $|\Omega|_{un} = p_I \cdot |\Omega|$, the addressable part $|\Omega'|$ and the immune part in which no defects occur $|\Omega|_{im} = p_{im} \cdot |\Omega|$. The fit parameter v was estimated with Monte-Carlo simulations. Figure 3.11 shows the sizes of these subspaces for a ternary (3,14)-reflexive code depending on the

Fig. 3.11 Dependency of different code space subsets on V_T variability

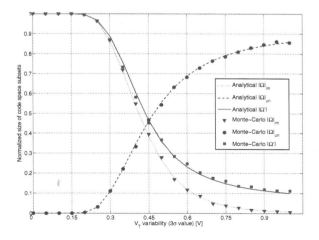

3σ-value of V_T. The monte-Carlo simulation confirms in the same figure the analytical results and gives the value 2.8 for the fit parameter v. The size of the addressable space $|\Omega'|$ drops quickly when 3σ reaches 0.4 V. At the same time, more patterns become uncovered. Interestingly, there are more addressable than immune patterns, because some defective patterns can be randomly addressed. This tendency increases for unreliable technologies, and around 10% of the original code space size can be randomly addressed under extreme conditions. The simulation of hot codes was not shown, because the result is similar, except for large defect probabilities: under these conditions the size of the addressable space goes faster towards 0 because the construction of hot codes imposes more constraints than the NRC.

The fact that the hot code space is under more constraints than the reflexive code space could be illustrated by investigating the value of v for different codes. It is important to emphasize the fact that v is a statistical quantity that gives the average number of elements in the code spaces mapped onto the pattern space under variability conditions through the relation "code covers pattern". Its value was estimated by matching the analytical results to the Monte-Carlo simulations. For hot codes, v has a value between 1.3 and 1.4 for all bases and lengths used in the Monte-Carlo simulations. On the other hand, v increases with the code length and the basis value for reflexive codes, which is illustrated in Table 3.2. Keeping in mind that v represents the average number of codes covering a given pattern under fluctuations, we conclude that reflexive codes are under less constraints than hot codes. This is confirmed by the fact that hot codes are a subset of tree codes, which are used as a seed for reflexive codes.

The sizing of the memory blocks (i.e., the size of contact groups in Fig. 3.1) and the number of V_T are dependent. As a matter of fact, Fig. 3.12 shows that increasing the number of V_T has two opposite effects: on one hand, it enables the addressing of more wires with the same code length; on the other hand, it makes the transistors more vulnerable to defects and increases the number of lost codes. A typical trade-off situation is illustrated in Fig. 3.12 with the ternary (3,9) and binary (2,12) hot codes (with $(n, k) = (3, 3)$ and (2,6) respectively) yielding almost the same number of addressable nanowires for 3σ around 0.4 V. The first one saves area because it has shorter codes, whereas the second one is technologically easier to realize (only 2 different V_T). The use of the ternary decoder is recommended for reliable technologies (insuring less area and more codes), but when the

Table 3.2 Estimated value of v for NRC with different code lengths M and bases n

M	6	8	10	12	14	16	18	20	22
$n = 2$	1.8	1.8	1.8	1.8	1.8	2.2	2.3	2.1	2.5
$n = 3$	1.9	2.2	2.2	2.5	2.8	2.8	3.0	–	–
$n = 4$	1.8	2.1	2.2	2.8	–	–	–	–	–
$n = 5$	2.1	2.2	2.7	–	–	–	–	–	–

Code space sizes $\gg 1$ Mbit identified by –

Fig. 3.12 Number of
addressable nanowires for
different hot codes

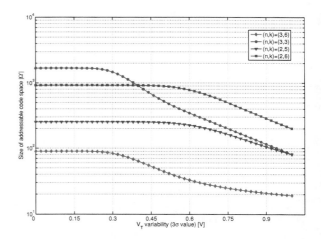

technology becomes more unreliable, there is a trade-off between the area-saving
and the easier fabrication process.

We also investigated the impact of the placement of V_A between two successive
$V_{T,i}$. We postulated in an intuitive way that V_A has to be the median of $V_{T,i}$ and
$V_{T,i+1}$. While modeling the defects, we allowed V_A to translate by a small value υ.
When υ increases (i.e., V_A moves towards $V_{T,i}$), then the probability of a flip-up
increases and that of a flip-down decreases. The opposite happens when V_A moves
towards $V_{T,i+1}$. The normalized number of addressable nanowires has been plotted in
Fig. 3.13. For unreliable devices with $3\sigma > 0.1$ V, the optimal position of V_A is
slightly shifted from the middle of $V_{T,i}$ and $V_{T,i+1}$ towards $V_{T,i+1}$ by a few tens of mV.
While reliable devices show a plateau around the optimal value of υ and necessitate
no accurate calibration of V_A, the circuit designer has to calibrate the applied

Fig. 3.13 Impact of the
value of V_A on the number of
addressable nanowires

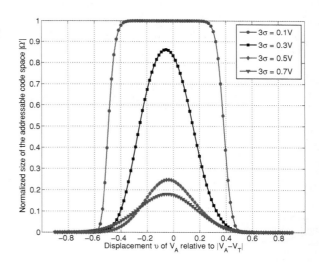

voltages V_A in a precise way when the transistors are not reliable; otherwise a certain loss in the number of addressable nanowires has to be taken into account. This circuit level issue has to be considered for either hot or n-ary reflected codes (both codes showed a similar behavior). In technologies with a 3σ-value below 100 mV, no calibration will be needed.

3.3.9 Simulations of the Effective Memory Area

3.3.9.1 Top-Down Approach: GAA Decoder Based Memories

We considered the *gate-all-around* (*GAA*) SiNW technology, explained in [30], since it represents a good candidate for crossbar decoders, then we explored the memory effective capacity/area design space and we performed a simulation of the design space (Fig. 3.14). The processes based on 2 and 4 V_T were considered for both hot and n-ary reflected codes. As Fig. 3.14 shows, the hot code generally reduces the decoder area, and consequently the memory area, because with the same code length, it is possible to address a larger code space. For instance, by using the same technology with 4 V_T to fabricate a 4 kb-memory, the NRC has an area overhead of $\sim 10\%$ compared to the hot code. The use of a simpler technology with 2 V_T implies an area overhead of $\sim 24\%$ for the same memory size. The area saving can be performed at either the technology or system level. It is worth to notice that the area saving is more significant for small memory sizes (typically less than 0.1 Mb), because the memory area for large memories is dominated by the area of the programmable array. The programmable array can be split into smaller blocks defined by the size of the contact groups (Fig. 3.1a) in order to reach the optimal size.

Fig. 3.14 Memory area/ capacity design space for different decoders

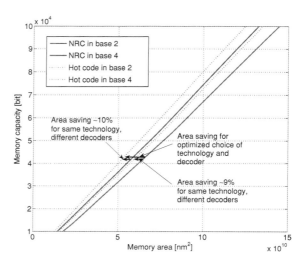

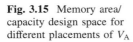
Fig. 3.15 Memory area/
capacity design space for
different placements of V_A

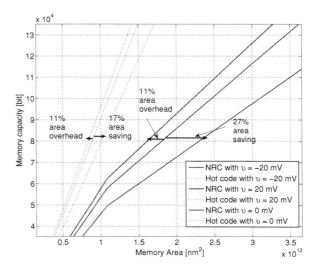

Since the placement of V_A between the successive V_T's is a critical design aspect, we investigated its impact on the overall memory area/capacity design space. In Fig. 3.15 we assumed a less reliable technology with 3σ-value around 1.2 V. If V_A is simply placed at the median of each two successive $V_{T,i}$ and no care is taken to place it at the optimal position (i.e., $\upsilon = -20$ mV), then the memory has an area overhead of around 11%. On the other hand, even at the suboptimal position $(\upsilon = 0)$, the area saving compared to the position corresponding to $\upsilon = 20$ mV is 17% for a hot decoder and 27% for a n-ary reflected decoder. These remarks confirm the importance of accurately controlling V_A in crossbar memories.

3.3.9.2 Bottom-Up Approach: Axial Decoder Based Memories

Multi-valued logic decoding can be used also within other decoder technologies. As an example of bottom-up approaches, we chose the axial decoders. The yield of axial decoders was investigated under different scenarios, by changing the process variability σ and the code type. The parameters used to estimated the memory area and density were given in [1] and summarized in Sect. 3.3.7. The particularity of decoders in this bottom-up approach is that they assign addresses randomly. The size of the contact group is set to the smallest value allowed in the considered technology in order to maximize the efficiency of the random address assignment [1].

We considered memory blocks with 32 kB raw density, where the raw density designates the memory density in the error-free case. We first set σ to a very low value (below 10 mV), so that the defects have a negligible impact on the bit area. The results are plot in Fig. 3.16, using a reflexive code with $n = 2$ and 3.

Fig. 3.16 Effective bit area versus code length for binary and ternary reflexive codes and $C_{\mathrm{raw}} = 32\,\mathrm{kB}$

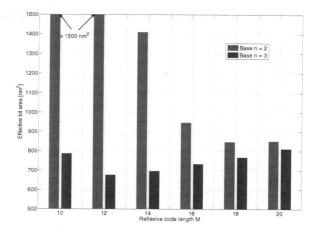

The results for hot codes have a similar qualitative behavior. For $n = 2$, the effective bit area is large for short codes because small code spaces do not insure enough unique codes. The randomness in the bottom-up decoding schemes necessitates a code space that is large enough to insure unique addressing. On the other hand, long codes increase the size of the decoder and the cost in terms of area cancels the gain in terms of unique codes. Between these two regions, an optimal code length exists for which the effective bit area is minimal. When we use a larger number of logic levels, the code space for short codes is already large enough to insure the unique addressing with random decoders. Thus, the optimal code length becomes smaller for larger n.

The same simulation described above was performed with a higher variability. The normalized results with the binary hot code are plotted in Fig. 3.17 for low ($3\sigma = 30$ mV) and larger ($3\sigma = 300$ mV) levels of variability. Each one of the two

Fig. 3.17 Effective bit area normalized to the minimal value versus binary reflexive code length for low and high variability and $C_{\mathrm{raw}} = 32\,\mathrm{kB}$

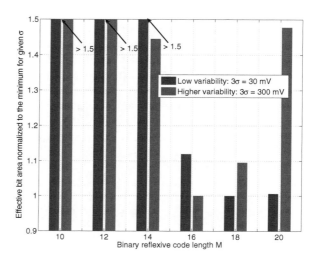

considered cases was normalized to its own minimal value, in order to illustrate how the optimal code length varies with variability. The decrease in effective bit area does not happen in a uniform way for all codes. Consequently, the position of the optimal code is shifted depending on σ. For instance, the optimal binary reflexive code has $M = 18$ for low σ, whereas the optimum is shifted towards $M = 16$ for the same code when σ becomes large. Designing the circuit with an overestimated reliability would result in about 10% larger effective bit area. This result shows that optimizing the choice of the decoder strongly depends on the estimated reliability level.

3.3.9.3 Summary

The benefits of using multi-valued logic to design bottom-up decoders is summarized in Table 3.3. Among the decoders presented in Sect. 3.2.1, the radial decoder would need several oxide shell thicknesses and the random contact decoder would need more than one level of conduction in order to be extended to n-ary logic. These features are not inherent to the decoders as presented in [2] and [4]; thus they cannot be extended to multi-level logic. On the contrary, it is possible to assume more than 2 levels of doping for the axial decoder and more than 1 oxide thickness for the mask-based decoder in order to perform a multi-valued logic addressing without altering the underlying decoding paradigm. Consequently, only these two decoders were extended to multi-valued logic addressing. The bottom-up approaches promise a high effective density under technological assumptions that are still to be validated. The use of ternary logic in 32 kB raw area memories saves area up to 20.2% for memories with axial decoder, and up to 11.8% for memories with masked-based decoder.

On the other hand memories using GAA decoders have a lithography-dependant pitch; and they consequently compete with the lithography-based memories. The crossbar architecture is the main reason for the smaller bit area when compared to DRAM or Flash memories. The larger the raw memory density, the better is the yield of the crossbar memory with GAA decoder (see Table 3.4). For high raw memory density, the ultimate limit is defined by the square of the minimal features (poly pitch) which is almost the same limit for Flash memories.

The area saving shown in the previous simulations is due to the decoder part, since no defects were assumed in the programmable part, where the information is

Table 3.3 Yield of different bottom-up decoders in terms of area per working bit (nm^2) at the technology node 45 nm

Raw size (kB)	Base	Axial decoder	Mask-based
8	2	1576	622
8	3	1196	550
8	Δ	24.1%	11.5%
32	2	846	423
32	3	676	373
32	Δ	20.2%	11.8%

Table 3.4 Yield of different top-down decoders in terms of area per working bit (nm^2) at the technology node 45 nm

Raw size (kB)	Base	GAA decoder	DRAM	Flash
8	2	9055	17690	9269
8	3	8651	17690	9269
8	Δ	4.5%	N/A	N/A
32	2	8153	16391	8355
32	3	7870	16391	8355
32	Δ	3.5%	N/A	N/A

Table 3.5 Decoder area in percentage of circuit area at the technology node 45 nm for different decoder types

Raw size (kB)	Base	Axial (%)	Mask-based (%)	Radial	Random cont.	GAA (%)
8	2	49.8	45.0	41.7%	45.1%	14.4
8	3	41.2	37.3	N/A	N/A	9.8
32	2	45.8	36.4	34.4%	38.3%	9.0
32	3	36.5	23.9	N/A	N/A	4.4

stored. The efficiency of using a higher logic level is related to the percentage of the decoder area compared to the whole memory; which is given in Table 3.5. For 32 kB raw memories, it ranges between 20 and 50% for bottom-up approaches and it is less than 10% for memories with GAA decoders because the programmable area is dominating. For smaller memories (8 kB raw density), the decoder part is larger in percentage because the decoder area scales logarithmically with the memory size.

3.4 The MSPT Decoder

In Chap. 2, the multi-spacer technique was demonstrated as a possible future technology for highly dense nanowire crossbars with a sub-photolithographic pitch. By enhancing the fabrication technique, it is possible to account for the decoder during the nanowire fabrication phase, as explained in Sect. 2.7.4. In this section, the design aspects related to the MSPT-based decoder concept are investigated. The focus will be on the optimization of the code type in order to simplify the fabrication complexity on the one hand, and to reduce the circuit variability on the other hand. The multi-valued logic codes defined in the previous section are used and optimized for the MSPT decoder.

3.4.1 Design of the Decoder

The MSPT decoder fabrication technique introduced in Sect. 2.7.4 yields a decoder operation identical to the description in Sect. 3.1. However, the layout differs because the nanowires lie within parallel caves having a symmetry axis going through their central axis as depicted in Fig. 2.27. The unique addressing of every nanowire in a half cave insures the unique addressing of every nanowire in the whole array. We will therefore consider only half caves in this study.

Every half cave contains N nanowires having M doping regions each. Recall that the pattern and doping profiles are the distribution of the threshold voltages and dopant concentrations among the M doping regions respectively. Let $\mathbf{P}_i = \begin{bmatrix} P_i^0 \dots P_i^{M-1} \end{bmatrix}$ and $\mathbf{D}_i = \begin{bmatrix} D_i^0 \dots D_i^{M-1} \end{bmatrix}$ be the pattern and doping profile of the nanowire i respectively. For the considered technique, whenever a nanowire i is patterned by receiving a doping dose, all nanowires $k = 0, \dots, i-1$ receive the same doping dose simultaneously. Consequently, the doping profile of a nanowire i depends not only on its own doping dose but also on all doping doses received by the nanowires $k = i+1, \dots, M-1$. We therefore need to determine the analytical multivariable application that links P_i^j and D_i^j for $i = 0, \dots, N-1$ and $j = 0, \dots, M-1$, in order to specify whether we can find a set of doping profiles that results in a given set of patterns.

Assuming that a set of doping profiles exists for any set of patterns, it is possible to optimize the choice of patterns according to different cost functions. We consider first the impact of this decoding technique on the fabrication cost. The nanowire profile implies a certain number of lithography/doping steps per nanowire, ϕ_i for $i = 0, \dots, N-1$. From the fabrication point of view, it is of the highest importance to reduce the total number of lithography/doping steps, i.e., $\sum \phi_i$. We therefore need to establish the link between P_i^j and ϕ_i in order to minimize $\sum \phi_i$.

Then, we consider the impact of this decoding technique on the circuit yield by analyzing the variability of the decoder. Every doping region j of the nanowire i, referred to as region (i,j), receives successive doping doses bit by bit. With every additional doping dose, the variability of region (i,j), quantified as the standard deviation of the threshold voltage of this region Σ_i^j, accordingly increases. It is therefore desirable to establish the link between P_i^j and Σ_i^j and to optimize the choice of the patterns in order to minimize the variability.

In the following section, we will derive the analytical mapping between patterns and doping profiles. Then, we will define the cost functions related to the fabrication complexity and to the threshold voltage variability. These cost functions will be minimized by choosing the best code type for the decoder.

3.4.2 Problem Formulation of MSPT-Based Nanowire Decoder

In this section we provide an abstract description of the decoder part of the nanowire array in a single half cave. The matrices defined below describe the most relevant design and fabrication aspects. Using these definitions, we will derive the cost functions of the fabrication complexity and the circuit variability. Further, we assume a multi-valued logic addressing with n values.

Definition 8 The *pattern matrix* **P** is an N × M matrix in $\{0, \ldots, n-1\}^{N \times M}$ representing the patterns of N nanowires within a half cave, where every nanowire has M doping regions.

We assume that the N nanowires within every half cave are patterned and have M doping regions each. The pattern corresponds to a set of M V_T's having any one of the n possible values $V_T(0), \ldots, V_T(n-1)$. The patterns are represented by the discrete values $0, \ldots, n-1$, which correspond to the ordered discretization of the threshold voltages $V_T(0), \ldots, V_T(n-1)$. Consequently, the set of patterns on the N nanowires forming one half cave can be represented by an N × M-matrix in $\{0, \ldots, n-1\}^{N \times M}$.

Definition 9 The *final doping matrix* **D** is an N × M matrix in $\mathbb{R}^{N \times M}$ representing the doping level distribution along the N nanowires within a half cave after the definition of the whole array.

Every V_T needs a unique doping level N_D fixed by the device physics and geometry [31]. Consequently, the pattern matrix, that is uniquely mapped onto a set of V_T's, defines a unique final doping matrix.

Example 11 For $n = 3, N = 3$ and $M = 4$, we assume that V_T can have the values 0.1 V, 0.3 V and 0.5 V corresponding to the digits 0, 1 and 2 and to the doping levels 2, 4 and 9 $\times 10^{18}$ cm^{-3}. The patterns are represented with the first N code words of the n-ary tree code. With **V** the matrix covering all V_T's, we obtain:

$$\mathbf{P} = \begin{bmatrix} 0 & 1 & 2 & 1 \\ 0 & 2 & 2 & 0 \\ 1 & 0 & 1 & 2 \end{bmatrix} \mathbf{V} = \begin{bmatrix} 1 & 3 & 5 & 3 \\ 1 & 5 & 5 & 1 \\ 3 & 1 & 3 & 5 \end{bmatrix} \cdot 0.1\,\text{V} \quad \mathbf{D} = \begin{bmatrix} 2 & 4 & 9 & 4 \\ 2 & 9 & 9 & 2 \\ 4 & 2 & 4 & 9 \end{bmatrix} \cdot \frac{10^{18}}{\text{cm}^3}$$

□

Proposition 5 *A non-linear bijective application h maps* **P** *onto* **D** *as follows:* $D_i^j = h(P_i^j) \, \forall \, i, j$

Proof The mapping between digits of the patterns and V_T is a discrete ordering, which is a bijective application g. The mapping between V_T's and N_D's is a monotonic non-linear function f, which is also a a bijection. The interested reader is invited to look into [31] to obtain the exact expression of f. Since h is a composition of f and g, it is bijective as well.   □

Definition 10 The *step doping matrix* **S** is an N × M matrix in $\mathbb{R}^{N \times M}$ representing the additional doping levels after every lithography/doping step.

There is a lithography/doping procedure that follows the definition of every one of the N nanowires. Every procedure $i = 0, \ldots, n-1$ is characterized by M doping levels $\left[N_{D,i}^0, \ldots, N_{D,i}^{M-1}\right]$ along the M doping regions of the nanowires. The set of M doping levels in the N steps can be represented by the matrix $\mathbf{S}$ in $\mathbb{R}^{N \times M}$.

Proposition 6 *A multi-linear application maps the elements of* $\mathbf{S}$ *onto those of* $\mathbf{D}$ *as follows:* $D_i^j = \sum_{k=i}^{N-1} S_k^j$

Proof Every nanowire j that is defined, is subsequently patterned by means of doping doses $\left[S_j^0, \ldots, S_j^{M-1}\right]$. Any nanowire i defined before the nanowire $j(i<j)$ receives the same dose simultaneously. Thus, the doping level of the nanowire i is the sum of all the levels defined in the steps $i, \ldots, N - 1$ following the definition of the nanowire i. □

Example 12 The following step and final doping matrices verify the property stated in Proposition 6. Negative and positive doping levels correspond to the doses with n- and p-type dopants respectively:

$$\mathbf{D} = \begin{bmatrix} 2 & 4 & 9 & 4 \\ 2 & 9 & 9 & 2 \\ 4 & 2 & 4 & 9 \end{bmatrix} \cdot \frac{10^{18}}{cm^3} \quad \mathbf{S} = \begin{bmatrix} 0 & -5 & 0 & 2 \\ -2 & 7 & 5 & -7 \\ 4 & 2 & 4 & 9 \end{bmatrix} \cdot \frac{10^{18}}{cm^3}$$

□

Definition 11 The *technology complexity* is quantified by Φ representing the total number of additional lithography/doping steps needed to pattern the nanowires.

Every row in $\mathbf{S}$ $(\mathbf{S}_i = \left[S_i^0 \ldots S_i^{M-1}\right], i = 0, \ldots, N - 1)$ represents the doping doses used in a single step doping procedure. The number of unequal non-zero elements in $\mathbf{S}_i$ represents the number of different doses used at this doping step. The more doping doses, the more lithography steps are needed and the more complex is the fabrication. Let $\phi_i (i = 0, \ldots, N - 1)$ be the number of unequal non-zero elements in $\mathbf{S}_i$, then the total number of lithography/doping steps is $\Phi = \sum_i \phi_i$.

Example 13 For $\mathbf{S}$ given in Example 12, we have: $\phi_1 = 2, \phi_2 = 4$ and $\phi_3 = 3$. Then, $\Phi = 9$ holds. □

Definition 12 The *decoder variability* is quantified by a $N \times M$ matrix Σ, describing the standard deviation of the threshold voltages in every doping region in the decoder of a half cave.

Every doping operation yields a V_T with a given variability σ_T, measured as the standard deviation. In the proposed technique every doping region is doped at most N times (Proposition 6). We expect the variability to increase with increasing number of doping operations. The number of times a doping regions (i, j) receives a doping dose decreases with increasing i and increasing number of zero-elements in the column j of $\mathbf{S}$. Let v_i^j be this number, then $v_i^j = \sum_{k=i...N-1} (1 - \delta(S_k^j))$, where $\delta(x)$ is the *Kronecker delta function*: $\delta(x) = 1 \Leftrightarrow x = 0$, otherwise $\delta(x) = 0$. Doping operations are assumed to be stochastically independent. The addition of two independent stochastic variables with standard deviations σ_1 and σ_2 respectively yields a stochastic variable with the standard deviation $\sigma_0 = \sqrt{\sigma_1^2 + \sigma_2^2}$. Therefore, if we define Σ as the N $\times$ M-matrix describing the variability of the decoder by setting Σ_i^j to the square of the standard deviation of the doping region (i,j), we obtain: $\Sigma_i^j = \sigma_T^2 \cdot v_i^j$.

Example 14 For $\mathbf{S}$ given in Example 12, we have:

$$\mathbf{S} = \begin{bmatrix} 0 & -5 & 0 & 2 \\ -2 & 7 & 5 & -7 \\ 4 & 2 & 4 & 9 \end{bmatrix} \cdot \frac{10^{18}}{\text{cm}^3} \quad \Sigma = \begin{bmatrix} 2 & 3 & 2 & 3 \\ 2 & 2 & 2 & 2 \\ 1 & 1 & 1 & 1 \end{bmatrix} \cdot \sigma_T^2$$

$\square$

Proposition 7 *Optimizing the decoder fabrication complexity consists in finding the best pattern $\mathbf{P}$ that minimizes Φ. Optimizing the decoder reliability consists in finding the best pattern $\mathbf{P}$ that minimizes $\|\Sigma\|_1$, where $\|\Sigma\|_1$ is the sum of all elements of Σ, known as its entrywise 1-norm.*

Proof This follows directly from Definitions 11 and 12. $\square$

3.4.3 Optimizing Nanowire Codes

In Sect. 3.2.2 we reviewed two main types of codes that have been used to uniquely address the nanowires in any logic with n values: hot codes and tree codes. We also reviewed the properties of the Gray code, which is an arrangement of the tree code that sets the number of transitions between two successive code words to 1.

In order to insure a unique addressing, tree codes have been used in a reflexive form, i.e., by appending to every code word its n-complement. We will use the reflexive form of the tree code and Gray code without any explicit reference. The length of the whole code word–including the reflected part–is M. The first two columns in Table 3.6 represent the binary tree and Gray codes (in their reflected form) with $M = 4$. Notice that the reflection multiplies the number of transitions between successive code words by 2.

In the following, we will prove that the Gray code is the optimal arrangement of the tree code with respect to the defined cost functions. Then, we will investigate the opportunity of arranging the hot codes in a similar fashion to Gray codes in order to optimize the costs of the decoders designed with hot codes.

Table 3.6 Example of binary codes with the length $M = 4$: (reflected) tree code (TC), (reflected) Gray code (GC), hot code (HC) and arranged hot code (AHC). τ represents the number of transitions from a code word to the following. Notice that the reflection doubles the number of transitions

TC		GC		HC		AHC	
Code	τ	Code	τ	Code	τ	Code	τ
0011	2	0011	2	1100	2	1100	2
0110	4	0110	2	1010	2	1010	2
1001	2	1100	2	1001	4	1001	2
1100	–	1001	–	0110	2	0101	2
–	–	–	–	0101	2	0110	2
–	–	–	–	0011	–	0011	–

3.4.3.1 The Gray Code

Proposition 8 *Among all arrangements of tree codes, the Gray code minimizes the decoder cost in terms of variability* $\|\Sigma\|_1$.

Proof For $i = N - 1$, $S^j_{N-1} = D^j_{N-1} = h^{-1}(P^j_{N-1})$ is fixed by the pattern of the last nanowire, i.e., by P^j_{N-1}. Thus, $v^j_{N-1} = 1 - \delta(S^j_{N-1}) = 1$, because $S^j_{N-1} \neq 0 \,\forall j$, since every region receives a doping dose in order to define the pattern of V_T's of the last nanowire P^j_{N-1}. For $i \neq N - 1$, $v^j_i - v^j_{i+1} = 1 - \delta(S^j_i) = 1 - \delta(D^j_i - D^j_{i+1})$. This difference is 1 if $D^j_i \neq D^j_{i+1}$, i.e., $P^j_i \neq P^j_{i+1}$, and 0 if $P^j_i = P^j_{i+1}$. Then, v^j_i can only increase by steps of 1 or remain constant with decreasing i for a fixed j. It remains unchanged if and only if the pattern P^j_i remains unchanged.

Consequently, $\|\Sigma\|_1$ monotonically increases with increasing transitions in the pattern matrix **P** between every two successive rows. Given that the rows of **P** are the code words in the chosen code space, it is desirable to use the code that minimizes the number of transitions between successive code words in order to minimize $\|\Sigma\|_1$. This condition is fulfilled by the Gray code. □

Example 15 Instead of **P** given in Example 11, which includes a tree code sequence with the cost $\|\Sigma\|_1 = 22 \cdot \sigma_T^2$ (from Example 14), we use a sequence from the Gray code that avoids the forbidden transition in **P** $0220 \Rightarrow 1012$. Then we obtain $\|\Sigma\|_1 = 18 \cdot \sigma_T^2$:

$$\mathbf{P} = \begin{bmatrix} 0 & 1 & 2 & 1 \\ 0 & 2 & 2 & 0 \\ 1 & 2 & 1 & 0 \end{bmatrix} \mathbf{S} = \begin{bmatrix} 0 & -5 & 0 & 2 \\ -2 & 0 & 5 & 0 \\ 4 & 9 & 4 & 2 \end{bmatrix} \cdot \frac{10^{18}}{\text{cm}^3} \quad \Sigma = \begin{bmatrix} 2 & 2 & 2 & 2 \\ 2 & 1 & 2 & 1 \\ 1 & 1 & 1 & 1 \end{bmatrix} \cdot \sigma_T^2$$

□

Proposition 9 *Among all arrangements of tree codes, the Gray code minimizes the fabrication cost* Φ.

Proof   In a similar way to the proof of Proposition 8, we notice that the value of S_i^j is unequal to zero if there is a transition $P_i^j \Rightarrow P_{i+1}^j$ between two successive code words in **P** at the digit j. Then, any ϕ_i, and consequently Φ, increase with the number of transitions in **P**. Since the Gray code minimizes the number of transitions, then it is optimal with respect to Φ. $\square$

Example 16   The Gray code in Example 11 has a fabrication cost $\Phi = 9$ (Example 13). By using the Gray code in Example 15, the fabrication cost was reduced to $\Phi = 7$ ($\phi_1 = 2, \phi_2 = 2$ and $\phi_3 = 3$). $\square$

3.4.3.2 Arranged Hot Codes

The previous section demonstrated that the arrangement of the tree code words into a sequence that minimizes the number of transitions between every successive code words, defined a new code (the Gray code) that minimizes the decoder variability and the fabrication cost. We therefore considered the question whether the code words of hot codes can be arranged in a similar way to the Gray code, such that the number of transitions is minimized, and to assess the possible benefits of such codes, that we called *arranged hot code (AHC)*.

Since the number of digits with a given value in every hot code (M, k) is fixed, then the minimum number of transitions is 2. We used an exhaustive algorithm for most of the hot codes with a reasonable code space size ($\lesssim 100$) for nanowire arrays, and we found that the arrangement in a *Gray-code-fashion* always exists. An example is given in the last two columns of Table 3.6.

It is possible to show in a very similar way to Proposition 8 and 9 that, when an arrangement of a given hot code exists, in such a way that the number of transitions between every successive code words is minimized, then this arrangement is the optimal hot code with respect to $\|\Sigma\|_1$ and Φ compared to all possible arrangements of the same hot code. In the next section we will therefore assess the performance of the optimized versions of both tree and hot codes in terms of fabrication complexity and circuit costs.

3.4.4 Simulations of the Decoder

3.4.4.1 Simulation Platform

In order to assess the impact of the decoder design (meaning the choice of the code type) on the fabrication complexity and the circuit features, we performed a statistical analysis of a crossbar circuit (Fig. 3.1a). The function of the crossbar circuit was assumed to be a memory as in the simulations done in the previous section. The defects affecting the molecular switches or the phase change layer were not simulated, given the fact that only defects affecting the decoder are addressed. Then, only the defects happening at the decoder part due to the variability of the V_T's in the doping regions were considered.

We assumed the same number of caves in both layers forming the crossbar. Their number and the one of the nanowires in every half cave N was fixed according to the raw crosspoint density set to $D_{RAW} = 16$ kB. The number of contact groups per half cave was minimized with respect to the code type (code space size Ω and code length M) and geometry (lithography pitch P_l and nanowire pitch P_n). While Ω and M were used as simulation parameters, P_l was set to 32 nm and P_n to 10 nm. According to standard layout rules, the minimum width of every contact group had to be set to $1.5 \times P_l$. The maximum width of every contact group was limited by the width of Ω nanowires at most that can fit in every contact group.

The threshold voltages V_T's were assumed to be distributed within the range 0 to 1 V, in order to account for a maximum supply voltage of 1 V. The doping levels were estimated from V_T's by using the assumptions in [31] for the most common materials used in standard CMOS processes. The variability σ_T of V_T was set to 50 mV. A nanowire was assumed to be addressable if V_T at every doping region varies within a small range. In this way, the probability that a nanowire is addressed was calculated from the Gaussian distributions of V_T's with the standard deviations given by Σ. We accounted for nanowires that may be addressed by two adjacent contact groups, as explained in [1], and we removed them from the set of addressable nanowires. This gives the estimate for the yield of every cave Y. Consequently, the effective array density that denotes the number of working crosspoints can be estimated as: $D_{EFF} = D_{RAW} \cdot Y^2$.

The simulation platform is presented in a schematic way in Fig. 3.18.

Fig. 3.18 Statistical analysis platform

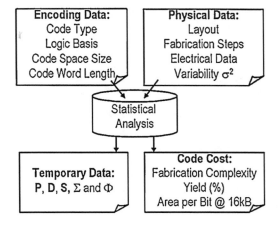

Fig. 3.19 Fabrication
complexity in terms of
number of additional steps for
different code types and logic
levels

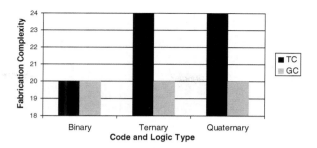

3.4.4.2 Simulation Results

We calculated the technology complexity Φ for different code and logic types. The results, plotted in Fig. 3.19 for $N = 10$ show that Φ is constant for all binary codes and equal to the double of the number of nanowires in a half cave. Higher logic level was suggested as a way to reduce the area overhead of the decoder as explained in Sect. 3.3.9. However, Fig. 3.19 shows that the higher logic level comes with some fabrication cost: 20% more steps for the tree code. For ternary and quaternary logic, the Gray code performs better than the tree code (17%) by completely canceling the fabrication complexity overhead.

The variability matrix was calculated for various types of binary codes. N was set to 20 and the plots in Fig. 3.20 show the variability level at every digit in the $N \times M$-matrix Σ, as square roots of elements of Σ normalized to σ_T. By comparing Fig. 3.20a, c and e, we see that the Gray code and its balanced version reduce the

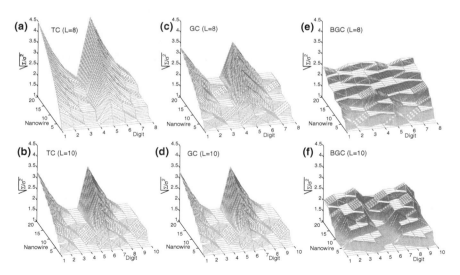

Fig. 3.20 Square root of elements of variability matrix Σ for different binary code types and lengths

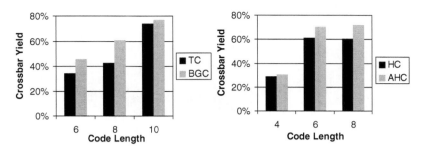

Fig. 3.21 Crossbar yield in terms of percentage of addressable crosspoints for different binary code types and lengths

variability level at every digit in comparison to the tree code. The balanced Gray code distributes the variability more evenly than the other codes. In this way, the average variability $\|\Sigma\|_1/(N \cdot M)$ could be reduced by 18%. Similar results were obtained for these codes with a higher logic level, as well as for hot codes and their arranged version. Next, we compared the distribution of the elements of Σ for a fixed code type and different code lengths (Fig. 3.20a, c, e vs. b, d, f). We noticed that longer codes have less digit transitions and help reduce the average variability.

The elements of Σ provide the inputs to estimate the crossbar yield, that we quantified as the effective crossbar density normalized by the raw crossbar density of 16 kB. The crossbar yield is plotted in Fig. 3.21 for various binary code types and lengths. The yield generally increases with increasing code length, until it reaches the maximum ($M \sim 10$ for TC/BGC and $M \sim 6$ for HC/AHC). The yield improvement of the tree code and the arranged hot codes, by increasing the code length from 6 to 10 and 4 to 8 respectively, is $\sim 40\%$. For a fixed code length, the optimized codes (i.e., BGC and AHC) perform better than their non-optimized versions (i.e., TC and HC respectively). For instance, the balanced Gray code yields 42% more than the tree code, and the arranged hot code 19% better than the hot code with the same length $M = 8$.

The dependency of the yield on the code length is explained by two factors: *i*) the variation with M of the percentage of nanowires at adjacent contact groups, which have to be removed from the set of addressable nanowires; and *ii*) the dependency of the variability on M. On one hand, by increasing M and keeping the code type fixed, the code space size increases and less contact groups are needed, so less nanowires are removed at adjacent contact groups and the yield increases. This effect saturates when the code space size is large enough to neglect the number of nanowires at adjacent contact groups. On the other hand, the average variability $\|\Sigma\|_1/(N \cdot M)$ decreases with increasing M, because longer codes have less digit transitions, as we showed previously. So, for a fixed code type, the yield, first, increases with increasing M, then it starts decreasing for larger M because the increasing number of digits cancels the benefits of decreasing variability of each digit taken separately. This decrease is just slightly seen for the hot code when M increases beyond 6; and for other code types, it starts appearing from $M \sim 10$.

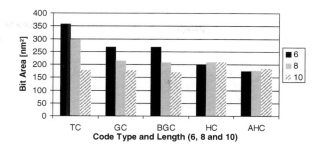

Fig. 3.22 Average area per functional bit for different binary code types and lengths

From the geometrical data, we estimated the crossbar area; then, by considering the effective density, we estimated the bit area for different code types and lengths (Fig. 3.22). For the tree code and its optimized versions (Gray and balanced Gray codes), the bit area decreases with increasing code length mainly due to the vanishing effect of adjacent contact regions, as explained previously. An area saving by 51% can be achieved by setting M to 10 instead of 6 for the tree code. The balanced Gray code yields a denser crossbar than the Gray code, which in turns yields better than the tree code; for instance crossbars with the balanced Gray code are 30% denser than those with the tree code (for $M = 8$). The hot and arranged hot codes yield the most dense crossbars with $M = 6$. For larger M, the bit area slightly start to increase, as the yield starts to decrease because of higher variability of longer codes, as explained previously. The arranged hot code performs better than the hot code with 13% less bit area for $M = 6$. Among all these codes, the smallest bit area is $169\,\mathrm{nm}^2$ for the balanced Gray code, followed by the arranged hot code with $175\,\mathrm{nm}^2$.

3.5 Discussions

This chapter represents an abstract formalism of nanowire encoding for different fabrication technologies. The idea of the chapter is to consider the nanowire part included in the decoder in an abstract way, as a sequence of digits in a code word. The impact of the physical variation of the underlying nanowire was modeled in an abstract way also, as a variation of the code word. The physical effect inducing the code word variation is assumed to be the variability of the threshold voltage. However, the operation of the crossbar is current-driven, i.e., the level of current indicates whether a crosspoint is addressed or not. It is therefore useful to extend the physical defect model to other sources of variation of the drive current, such as the nanowire cross-section dimension and the oxide thickness in the decoder part of the nanowires.

The purpose of the simulations presented in this chapter is to assess the impact of the decoder design (i.e., code type and decoder size) on yield and area. If a more accurate estimation of the memory yield is needed, then the defect model of the molecular switches has to be included in the model as well.

The predicted results are more reliable for nanowires with *reasonably small width and pitch*. It is difficult to accurately determine how small *reasonably small dimensions* are; but the results are expected to hold beyond 10-nm. If The nano-wire thickness is much less than 10 nm, then the number of dopants is expected to be too small to generate an accurate statistical distribution of the threshold voltage, unless a high σ_T and an approximative statistical description are accepted. If the nanowire pitch is too small, then every two adjacent nanowires are expected to have an electrostatical coupling and to mutually affect their threshold voltages, in a similar way to SOI double gate transistors [32].

A possible question that may arise with respect to the proposed approach is: how much is the benefit of using a computational model compared to Monte-Carlo simulations? Looking at the proposed models, it is clear that the most of them are explicit analytical expressions that can be computed instantaneously; and only some of those models describing hot codes are based on an iterative approach that is exhaustive in some cases necessitating all code words to be listed. Only these special cases are expensive in computation time, but they are still faster than Monte-Carlo simulation with many thousands of runs, given the fact that only reasonable code spaces with no more than a few hundreds of code words are needed for crossbars. In general, Monte-Carlo simulations are possible, but the utilization of the given computational models gives a better insight into the impact of every circuit or process parameter on the final result.

The use of MVL to encode nanowires showed non-negligible benefits in terms of area and yield, namely in dense crossbars with a small pitch between nanowires. The decoder, which represents the link between the crossbar and the CMOS part of the circuit, has to be designed in MVL as well, and needs to convert all signals from the binary level that is used in the CMOS part. It is consequently important to consider the costs of the proposed solution in terms of area overhead, power consumption and delay caused by the 2-to-n conversion.

The MSPT decoder is the first proposed digital decoder for MSPT-based nanowires. Its yield and area can be improved by optimizing the choice of the code type and length. However, it comes with a high cost in terms of fabrication complexity, i.e., number of photolithography/doping steps. The impact of this problem can be reduced by considering the advantage of the MSPT process to parallelize the processing within one single batch (by increasing the number of crossbars processed in parallel) and among different batches (by processing different wafers at the same time) as explained in Sect. 2.8 and Fig. 2.28. This decoder technology is not the only one, and it represents another solution besides the existing ones. The choice among different technologies will be made depending on the most relevant aspect for the final product: CMOS-compatibility, time-to-market, yield, etc.

3.6 Chapter Contributions and Summary

In this chapter, a novel approach to model nanowires and their defects in the decoder part of the crossbar is presented. The abstraction level gives a better insight into the impact of the decoder design parameters on the whole circuit. In this approach, the nanowires are modeled as a set of code words in a code space, and the variability affecting their physical parameters is modeled as errors affecting the code space.

This chapter proposes a method to construct a new set of codes that can be used to encode nanowires. The new codes are a generalization of hot and tree codes. Besides their ability to uniquely address the nanowires in a compact way, they offer interesting trad-off opportunities between code size and error probability if the variability is high. Another set of codes is obtained by arranging hot codes in an optimized way (like Gray codes), and shown to be compact and fault-tolerant. The existence of this set of codes has not been known and its use has not been considered in the past.

The decoder design has been given a novel direction in this chapter. Optimizing the decoder design is not only a matter of physical dimensions of the decoder, but it highly depend on the encoding scheme as well. Then, the decoder design can be driven by yield: there are different trade-off situations between the circuit area and yield depending on the code size and type used in the decoder.

This chapter explores for the first time the design aspects of the MSPT-decoder. It presents a novel approach to look at the problem across the fabrication and circuit design. The wide choice of codes spans the design space between the fabrication complexity and the overall circuit variability, and offers novel opportunities to simplify the fabrication, while keeping the variability of the circuit low enough.

The variability caused by the small nanowire dimensions is driving this chapter. The decoder design methodology presented here takes the variability into account, and tries to minimize the number of nanowires which are badly addressed; i.e., either not addressed by the decoder, or sharing the same address with other nanowires. The existence of such codes is demonstrated in this chapter, and the way to minimize their occurrence is the core of the proposed design methodology. The system needs to be provided with a method to detect their existence and to discard them. Such a system provides a decoder test procedure, which is the topic of the following chapter.

References

1. DeHon A, Lincoln P, Savage J (2003) Stochastic assembly of sublithographic nanoscale interfaces. IEEE Trans Nanotechnol 2(3):165–174
2. Savage JE, Rachlin E, DeHon A, Lieber CM, Wu Y (2006) Radial addressing of nanowires. ACM J Emerg Technol Comput Syst 2(2):129–154

3. Kuekes PJ, Williams RS (2001) Demultiplexer for a molecular wire crossbar network (MWCN DEMUX). US Patent 6,256,767, 2001
4. Hogg T, Chen Y, Kuekes P (2006) Assembling nanoscale circuits with randomized connections. IEEE Trans Nanotechnol 5(2):110–122
5. Gopalakrishnan K, Shenoy RS, Rettner C, King R, Zhang Y, Kurdi B, Bozano LD, Weslser JJ, Rothwell MB, Jurich M, Sanchez MI, Hernandez M, Rice PM, Risk WP, Wickramasinghe HK (2005) The micro to nano addressing block. In: IEEE Electron Devices Meeting, p. 19.4
6. Ben Jamaa MH, Atienza D, Moselund KE, Bouvet D, Ionescu AM, Leblebici Y, De Micheli G (2008) Variability-aware design of multi-level logic decoders for nanoscale crossbar memories. IEEE Trans Computer-Aided Des 27(11):2053–2067
7. Ben Jamaa MH, Leblebici Y, De Micheli G (2009) Decoding nanowire arrays fabricated with the multi-spacer patterning technique. In: Design Automation Conference (DAC), July 2009, San Francisco, California, USA
8. Luo Y, Collier CP, Jeppesen JO, Nielsen KA, DeIonno E, Ho G, Perkins J, Tseng H-R, Yamamoto T, Stoddart JF, Heath JR (2002) Two-dimensional molecular electronics circuits. J Chem Phys Phys Chem 3:519–525
9. DeHon A (2005) Design of programmable interconnect for sublithographic programmable logic arrays. In: Proceedings of the International Symposium on Field-Programmable Gate Arrays (FPGA), pp 127–137
10. Beckman R, Johnston-Halperin E, Luo Y, Green JE, Heath JR (2005) Bridging dimensions: demultiplexing ultrahigh density nanowire circuits. Science 310(5747):465–468
11. Anderson DA, Metze G (1995) Design of totally self-checking check circuits for m-out of-n codes. In: Twenty-Fifth International Symposium on Fault-Tolerant Computing Highlights from Twenty-Five Years, pp 244– 248, 27–30 June 1995
12. Gardner M (1972) The curious properties of the Gray code and how it can be used to solve puzzles. Sci Am 227:106–109
13. Gray F (1953) Pulse code communication. US Patent No. 2632058, 1953
14. Bhat GS, Savage CD (1996) Balanced Gray codes. Electron J Comb 3(1):R25
15. Smith KC (1981) The prospects for multivalued logic: a technology and applications view. IEEE Trans Comput 30(9):619–634
16. Smith KC (1988) Multiple valued logic: a tutorial and appreciation. Computer 21(4):17–27
17. Current K (1994) Current-mode CMOS multiple-valued logic circuits. IEEE J Solid-State Circuits 29(2):95–107
18. Ogawa K, Shibata T, Ohmi T, Takatsu M, Yokoyama N (1998) Multiple-input neuron MOS operational amplifier for voltage-mode multivalued full adders. IEEE Trans Circuits Systems II: Analog Digit Signal Process 45(9):1307–1311
19. Kencke D, Richart R, Garg S, Banerjee S (1998) A multilevel approach toward quadrupling the density of ash memory. IEEE Electron Device Lett 19(3):86–88
20. Mahapatra S, Ionescu AM (2005) Realization of multiple valued logic and memory by yybrid SETMOS architecture. IEEE Trans Nanotechnol 4(6):705–714
21. Miller D (1993) Multiple-valued logic design tools. In: Proceedings of the 23rd IEEE International Symposium on Multiple Valued Logic, Sacramento, California, USA, pp 2–11, 24–27 May 1993
22. Sasao T (1989) On the optimal design of multiple-valued PLAs. IEEE Trans Comput 38(4):582–592
23. Rudell R, Sangiovanni-Vincentelli A (1987) Multiple-valued minimization for PLA optimization. IEEE Trans Computer-Aided Des Integr Circuits Syst 6(5):727–750
24. Sasao T (1993) EXMIN2: A simplification algorithm for exclusive-OR-sumof- products expressions for multiple-valued-input two-valued-output functions. IEEE Trans Computer-Aided Des 12(5):621–632
25. Song N, Perkowski M (1996) Minimization of exclusive sum-of-products expressions for multiple-valued input, incompletely specified functions. IEEE Trans Computer-Aided Des Integr Circuits Syst 15(4):385–395

26. Files C, Perkowski M (2000) New multivalued functional decomposition algorithms based on MDDs. IEEE Trans Computer-Aided Des Integr Circuits Syst 19(9):1081–1086
27. Jiang Y, Brayton R (2000) Don't cares and multi-valued logic network minimization. In: IEEE/ACM International Conference on Computer Aided Design, 2000. ICCAD-2000, pp 520–525
28. Rachlin E (2006) Robust nanowire decoding. http://www.cs.brown.edu/publications/ theses/ masters/2006/eerac.pdf 2006
29. International technology roadmap for semiconductors (ITRS) (2007) http://www.itrs. net/reports.html. Tech. Rep., 2007
30. Moselund KE, Bouvet D, Ben Jamaa HH, Atienza D, Leblebici Y, De Micheli G, Ionescu AM (2008) Prospects for logic-on-a-wire. Microelectron Eng 85:1406–1409
31. Sze SM, Ng KK (2007) Physics of semiconductor devices. Wiley-Interscience, New Jersy
32. Cristoloveanu S (1995) Electrical characterization of silicon-on-insultaor materials and devices. Springer, Heidelberg

Chapter 4
Decoder Test

The decoder implements the task of linking the crossbar and the CMOS part of the circuit. The major challenge of the decoder is to bridge the different dimensions, given the fact that the crossbar may be defined on the sub-photolithographic scale, while the CMOS circuit is defined with state-of-the-art photolithography. It also has to guarantee a good reliability level by assigning a unique address to every nanowires, while keeping the area small and the fabrication simple enough. These constraints on the decoder have been addressed in Chap. 3; and the decoder design has been improved by optimizing the choice of the code and decoder dimensions.

The physical defects affecting the nanowires have been modeled at a high abstraction level as changes in the nanowire addresses. A defect can cause a change of the nanowire address such that the nanowire becomes unaddressable in the considered code space, or it shares the same address with another nanowire. In these cases, it is required that defective nanowire addresses are detected and discarded from the used set of addresses. This task can be performed by testing the decoder circuit.

Testing the decoder, in order to keep only defect-free parts of the code space, highly simplifies the test procedure of the whole crossbar circuit. This chapter proposes a test method that identifies the defective codes. The method quantifies the test quality, measured as the probability of test error, and it investigates the dependency of the test quality on the decoder design parameters.

This chapter is partly taken from [1] and it is organized the following way. First, the need to test the decoder alone in order to simplify the overall crossbar test is explained. Then, the current-based test methodology is introduced. The model assumptions on the variation of the current components are presented. The model is then applied with realistic values of the parameters, and simulations are performed to calculate the optimal test variables and to demonstrate the impact of the decoder design parameters on the test quality. Finally, the obtained results are discussed and a summary of the contributions of this part of the work is given.

M. H. Ben Jamaa, *Regular Nanofabrics in Emerging Technologies*,
Lecture Notes in Electrical Engineering, 82, DOI: 10.1007/978-94-007-0650-7_4,
© Springer Science+Business Media B.V. 2011

4.1 Necessity of Testing Crossbar Circuits

Without loss of generality, crossbar circuits considered in this part of the work implement a memory function. The reason behind this assumption is that the architecture of a crossbar memory is identical to the one a crossbar performing computation with passive elements (i.e., diodes). The READ and WRITE operations in the memory correspond respectively to output computation and switch configuration in the logic circuit. Testing the crossbar circuit performing either functions (memory or computation with passive elements) can be therefore performed with the same procedure. This chapter introduces first the problem of testing the whole circuit. Then, it focuses on testing the decoder part, which does not depend on the function of the circuit, i.e., whether it is memory or logic.

4.1.1 Operation of Crossbar Memories

Even though there are no complete memory systems based on the crossbar architecture yet, we believe that it will have the same architecture as CMOS memories [2], which is illustrated in Fig. 4.1a. Unlike conventional RAM, crossbar memories have two parts: a sub-lithographic part formed by the decoder and the memory array (Fig. 4.1b) and fabricated in the nanowire crossbar technology; and a lithographic part formed by the rest of the circuit and fabricated in CMOS technology.

The information is assumed to be stored in molecular switches grafted to every pair of crossing nanowires. In the on-state, the molecule is conducting (logic 1) and in the off-state, it is high resistive (logic 0). The WRITE operation is performed by, first selecting the bit to be written, and then, by applying a large positive or negative voltage at the pair of nanowires connected by the molecular

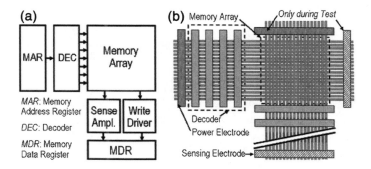

Fig. 4.1 Crossbar memory architecture. **a** RAM architecture at the functional level. **b** Sub-lithographic part of a crossbar memory

switch in order to set the molecular state, i.e., the bit value. On the other hand, the READ operation is current-based. In fact, if the molecule is in the off-state, then the nanowire in the lower level is almost floating [3] and no correct voltage level can be sensed. Consequently, the READ operation is performed by selecting the bit to be read, then by measuring the current through the sensing electrode (see Fig. 4.2a–d). Thus, the current-based READ operation in crossbar memories necessitates a thresholder as a part of the sense amplifier, in order to set the limit between the logic values 0 and 1, and to translate them into logic levels that can be stored in the memory data register.

4.1.2 Testing Complexity

Many sources of variability may cause errors in the sensed signal, such as the doping level and geometry variation. In particular, we focus on defects caused by the nanowires in the memory array and in the decoder. The doping level and geometry variation in the nanowire part forming the memory array induce a change in the resistance of the nanowires and a variation of the sensed current level. In the decoder part, these sources of variability can induce a drastic change in the on-resistance of the transistors forming the decoder by modifying their threshold voltage as analyzed in detail in the following paragraphs.

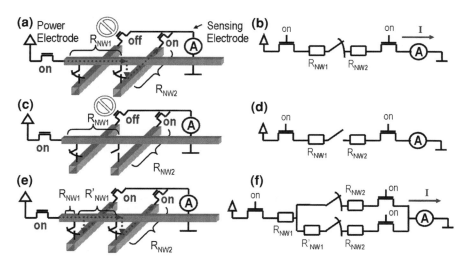

Fig. 4.2 Read operation in a memory with two bits. During correct operation, only one bit is addressed and read. **a** Correct reading of '1'. **b** Equivalent circuit for reading '1'. **c** Correct reading of '0'. **d** Equivalent circuit for reading '0'. **e** Defective reading of two bits. **f** Equivalent circuit for defective reading

The threshold voltage variation has been shown to cause defects in the decoder in such a way that by applying an address, any number of nanowires can be activated instead of one single nanowire. Fig. 4.2e, f show an example of defective addressing in the second nanowire layer. Thus, the sense amplifier reads the superposition of the information stored in more than one single bit. The thresholder cannot properly distinguish between the sensed signals resulting from the following cases: (i) one bit with the value 1, and (ii) the superposition of two bits whereby at least one of them has the value 1. In such a situation, the READ operation of the first bit yields a result depending on the state of the second bit, which causes *coupling faults* (*CFs*) in the memory [2]. Considering the fact that decoder defects typically make 2, 3 or more nanowires in each array active with the same address, the number of inter-dependant bits can be as large as 4–9 or even more, without necessarily having neighboring locations. This leads to the more critical *pattern sensitivity faults* (*PSFs*) [2].

In order to avoid complex and exhaustive PSF test procedures on the whole memory [4], it is attemptable to resolve the PSF caused by the decoder defects, before performing the conventional memory test. The thresholder can carry out this operation by checking the addresses of all nanowires in every layer (after separating them) and keeping only the addresses that activate one single nanowire. This procedure has a linear complexity with N, the number of nanowires in a layer (where N^2 is the number of bits in the memory). While it represents an additional testing step, this testing procedure, that we call *nanowire test*, resolves the necessity of an exhaustive PSF testing of the whole memory, whose complexity is exponential with N^2. However, we expect that the molecular switches will also induce PSFs. Since only neighboring molecules are likely to interact with each other, one can assume neighborhood patterns for the PSFs caused by the molecules. Therefore, simplified PSF procedures having a linear complexity with N^2 can be applied [4].

4.2   Testing Crossbar Circuits

This section presents an overview of a test method that can be applied to nanowire arrays. This is an exhaustive method used to illustrate the testing principle. More efficient pseudo-random techniques also exist. However, the focus here is only on the thresholder design and the test quality. The second part of this section is dedicated to the definition of the stochastic conditions of the test that optimize its quality.

4.2.1   Test Method

The nanowire testing is performed for every layer separately. Thus, we depicted a single nanowire layer with its additional test circuitry in Fig. 4.3. Besides the

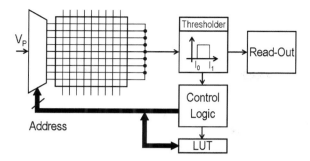

Fig. 4.3 Crossbar memory and testing unit: besides the memory array and the decoder, the system comprises a CMOS part formed by a thresholder that detects the bit state, a control unit that synchronizes the test operation, and a LUT that saves correct addresses

nanowire layer, the system comprises the interfacing circuit (decoder) and a CMOS part formed by a thresholder, a control unit and a LUT. The thresholder measures the output current and indicates whether a single nanowire is detected. The control unit regulates the execution of the testing phase; it can also control the execution of other functions, such as the reading and writing operations. The LUT is allocated to the considered nanowire layer, it stores the valid addresses activating a single nanowire each. Depending on the chosen granularity level, the thresholder, the control unit and the LUT can be used for one single nanowire layer, or for either layers defining a single nanowire array, or even for several nanowire arrays.

Normally the considered code testing phase comes into play only for the first-time test after manufacturing. However, if the defect rate is time-dependent, it is possible to consider running the code testing periodically in order to regularly update the table of defect-free nanowires. A correct testing can be performed by applying the following exhaustive procedure:

1. First, the vertical and horizontal layers forming the array are disconnected from each other, by setting the power (V_P) and sense (GND) electrodes of every layer at the same voltage, so that the voltage difference between both layers is large enough to program all the molecular switches into the off state.
2. Then, one of the two layers is considered first. The power electrode is set to a high voltage, while the decoder selects a specific code. If the code addresses one single nanowire, then a current flows and the code is saved in the LUT as potentially valid. If the pattern corresponding to the code is defective, then its nanowire is not conducting, and the sensing electrode detects a low current flow. Hence, the code is not saved in the LUT. If the defects make more than one nanowire controlled by the given code, then the sensing electrode detects a high current and the code is not saved in the LUT. This operation is repeated with all available codes and the LUT is filled up accordingly.
3. The analysis in Chap. 3 proved that a defective nanowire can be controlled by more than one code. In order to discard the nanowires undergoing this kind of

defects, a second run through all the codes saved in the LUT so far has to be undertaken. The first code in the LUT is applied at the decoder, while a high voltage is applied at the power electrode. If a current flows through the NW, then a flag is set on in the LUT to indicate that the code is definitely correct, i.e., it controls one single nanowire. Then, the corresponding nanowire is disconnected from the sensing electrode by applying the programming voltage at the power electrode. This step is repeated with all codes remaining in the LUT. Assuming that a nanowire has a defect that makes it controllable by two codes; then both codes are stored in the LUT, in step 2. After the first one is checked in step 3, the nanowire is disconnected from the sensing electrode. Consequently, when the second code is checked, no current flows; then the flag of this code is set off, meaning that the code is controlling a nanowire, which already has a code. Finally, only the codes having a positive flag in the LUT are kept, all the others can be removed from it. The remaining codes can be safely used during the normal operation of the crossbar array.

4. The same procedure is applied for the second layer and the correct codes are saved in a different LUT.

Filling-up every LUT needs N tests and updating it with the flags needs at most N tests, with N the number of nanowires in a single array. Thus, the cost of this exhaustive procedure is $\leq 2N$.

4.2.2 Test Requirements

The output of the nanowire layer (I_s) is sensed by the thresholder. We assume that the variability mainly affects the sub-lithographic part of the memory representing the nanowire array. This part is fabricated using an unreliable technology, unlike the rest of the circuit, defined on the lithography scale and assumed to be more robust. Thus, we consider that the thresholder, the control circuit or the LUT are defect-free. The thresholder senses I_s, it possibly amplifies it, then it compares I_s to two reference values (I_0 and I_1 with $I_0 < I_1$). If the sensed current is smaller than I_0, then no nanowire is addressed. If the sensed current is larger than I_1, then at least two nanowires are activated with the same address. If the sensed current is between the reference current levels, then only one nanowire is activated and the address is considered to be valid.

Previously, two effects resulting from the sub-lithographic nanoscale size of the nanowire layers have been highlighted: the randomness of the pattern allocation (Sect. 3.2.1) and the pattern variation due to the variability of the threshold voltages (Sect. 3.3.4). In the ideal case, every pattern is unique and by applying the code corresponding to a given unique pattern, the sensed current has a deterministic value (Fig. 4.4a). If we only consider the randomness in pattern allocation without considering the issue of V_T variability, then the sensed current can be modeled as a stochastic variable. The distribution can be considered discrete, since

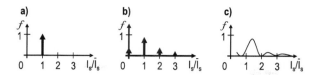

Fig. 4.4 Probability density function of sensed current. *Arrows* indicate Dirac pulses, i.e., discrete probability density function. **a** Deterministic addressing without random code allocation; **b** random code allocation without threshold voltage variability; **c** random code allocation with threshold voltage variability

without considering any variation, the values of V_T would be fixed (Fig. 4.4b). However, by additionally taking into account the variability of the threshold voltage, the sensed current can take any value and its distribution becomes continuous (Fig. 4.4c). In our approach, we assume a continuous distribution of the sensed current in order to take into account the variability of V_T.

The detection of correct addresses becomes more challenging under these conditions. The correct design of the thresholder circuit necessitates the use of the optimal values of I_0 and I_1. These are obtained by maximizing the probability that a correct address is detected (P_1: the conditional probability that I_s is between I_0 and I_1 given that only one nanowire is activated), and the probabilities that a defective address is identified as such (P_0 and P_2: the conditional probability that I_s is below I_0 or beyond I_1 given that no nanowire or more than one single nanowire are activated, respectively):

$$\begin{cases} P_0 = \Pr\{(I_s \leq I_0) \text{ given that no nanowire is addressed}\} \\ P_1 = \Pr\{(I_0 < I_s < I_1) \text{ given that 1 nanowire is addressed}\} \\ P_2 = \Pr\{(I_1 \leq I_s) \text{ given that} \geq 2 \text{ nanowires are addressed}\} \end{cases} \quad (4.1)$$

Then, the probability that all three events happen simultaneously is given by: $P_0 \times P_1 \times P_2$ (assuming that the considered events are independent). Consequently, we can define metrics for the test result, represented by the error probability of this test procedure:

$$\epsilon = 1 - P_0 \times P_1 \times P_2 \quad (4.2)$$

The purpose of this part of the work is to design the thresholder in order to obtain the best test result (i.e. with the smallest ϵ). The test result does not give any explicit information about the number of addressable nanowires; it rather indicates how good the estimation of the addressability of these nanowires is. In the rest of the chapter, we derive the analytical expressions of P_0, P_1 and P_2. We first model the stochastic distribution of the sensed signal under variability conditions; then we optimize the choice of the reference currents I_0 and I_1 in order to minimize the test error ϵ.

4.3 Perturbative Current Model

From the mathematical point of view, it should be possible to derive the *exact* distribution of the sensed current by considering the parameters and the $I-V$ characteristics of the circuit formed by the SiNWs. However, even the most basic $I-V$ characteristics of the transistors are not linear; thus making the derivation of an analytical expression of the distribution of the sensed current difficult. If advanced electrical effects for highly scaled transistors are taken into account, then the characteristics may even become non-analytical, which makes the derivation of the analytical expression of the distribution of the useful current impossible. We introduce in this section the fundamental approach that enables the modeling of the sensed current. We focus here on correctly addressed nanowires and we consider the impact of the variability of the $V_{T,i}$'s on the distribution of the current through these nanowires.

During the code testing phase, every nanowire is disconnected from the crossing nanowires as explained in Sect. 4.2.1. It can be modeled as a wire connecting the power electrode to the sensing electrode and formed by two parts (see Fig. 4.5): the decoder part that is a series of M pass transistors, and the memory part. Since the memory part is disconnected from the second layer of nanowires, it can be modeled as a resistive load R_M and the resistance of the decoder part can be omitted because the transistors are separated by one single lithography pitch, which is generally highly doped and low resistive (as for drain and source diffusion regions in a MOSFET).

The transistors in the decoder part of the nanowire are SiNWFETs. Their model is expected to include more scaling and coupling effects than the usual model for bulk MOSFET. We model the devices in this section in a general way as a *black-box* representing voltage-controlled current sources, i.e.: $I = f(V_{DS}, V_{GS}, V_T)$ where I is the drain-source current, V_{DS}, V_{GS} and V_T are respectively the drain-to-source, gate-to-source and threshold voltages. The decoder design approaches explained in Sect. 3.2.1 are based in the simplest case, namely, on two types of transistors having two different V_T's ($V_{T,\mathrm{Ref0}}$ and $V_{T,\mathrm{Ref1}}$ such that $V_{T,\mathrm{Ref0}} < V_{T,\mathrm{Ref1}}$, and we define $\Delta V_T = V_{T,\mathrm{Ref1}} - V_{T,\mathrm{Ref0}}$). We further assume that changing the

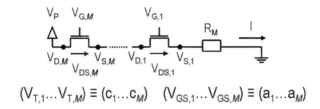

Fig. 4.5 Electrical parameters of a biased nanowire under test: the decoder part is represented by M transistors in series, and the memory part is represented by a resistance R_M. Notice that the perpendicular nanowire layer is disconnected from the nanowire under test

transistor type only changes its threshold voltage but not the current output function f. As a matter of fact, the transistors in the decoder part have the same p- or n-type even though they may have different threshold voltages. A high variation of f would be noticeable only if the transistors would have different polarities (p- and n-type), which would imply a remarkably different electron and hole mobility and, consequently, different expressions for f.

We consider a nanowire as modeled in Fig. 4.5, where a sequence of control voltages $(V_{GS,1}, \ldots, V_{GS,M})$ corresponding to its pattern are applied. Thus, this sequence switches every transistor on and the M conducting transistors generate a current flow I through the nanowire. Every variation of V_T results in a variation of the current through the nanowire, which can be noted the following way:

$$I = I^{OP} + \delta I \tag{4.3}$$

In this context, the signal I is linearized around the *operating point* (*OP*) and divided into a *large* I^{OP} and a *small signal* δI. This approach is widely used in circuit and network theory and in sensitivity analysis [5]. It gives a powerful way to simplify the study of complex circuits in an accurate way. The large signal needs a precise device model and can be estimated with a SPICE simulator. A hand estimation of I^{OP} is generally not possible because only numerical solutions can be found. On the contrary, the small signal is easier to calculate by linearizing all the equations describing the circuit around the OP.

In the following, we derive the general expression of δI. We assume that the variation of $V_T(\delta V_T)$ results in a variation of the circuit-internal variables: I and V_{DS}, but not in the fixed variables, such as, the voltage of the sensing electrode set as voltage reference, the power supply (V_P) and the gate voltage (V_G), both with respect to the reference.

We denote by i the transistor index $(i = 1 \ldots M)$. Then, the serial connection of the transistors results in the following gate-source voltage drop at transistor i:

$$V_{GS,i} = V_G - R_M \cdot I - \sum_{j=1 \cdots i-1} V_{DS,j} \tag{4.4}$$

Since $V_{GS,1} = V_G - R_M \cdot I$, we use the following convention: $\sum_{j=1 \cdots 0} V_{DS,j} = 0$. The Eq. 4.4 shows that V_{GS} can vary with V_T despite the fact that V_G is constant. The other equations describing the circuit of Fig. 4.5 are:

$$V_P = \sum_{i=1 \cdots M} V_{DS,i} + I \cdot R_M \tag{4.5}$$

$$I = f(V_{DS,i}, V_{GS,i}, V_{T,i}) \forall i \tag{4.6}$$

The linearization of the previous equations yields respectively to:

$$\delta V_{GS,i} = -R_M \cdot \delta I - \sum_{j=1 \cdots i-1} \delta V_{DS,j} \forall i \tag{4.7}$$

$$0 = \sum_{i=1\cdots M} \delta V_{DS,i} + R_M \cdot \delta I \tag{4.8}$$

$$\delta I = \frac{\partial f_i}{\partial V_{DS,i}} \Big|^{OP} \delta V_{DS,i} + \frac{\partial f_i}{\partial V_{GS,i}} \Big|^{OP} \delta V_{GS,i} + \frac{\partial f_i}{\partial V_{T,i}} \Big|^{OP} \delta V_{T,i} \forall i \tag{4.9}$$

By substituting Eqs. 4.7 and 4.8 in Eq. 4.9, for every $i = 1 \cdots M$, the following linear relation is obtained:

$$\delta \mathbf{V}_{DS} = \mathbf{A}^{-1} \cdot \mathbf{B} \cdot \delta \mathbf{V}_T \tag{4.10}$$

where the variational vectors are: $\delta \mathbf{V}_{DS} = [\delta V_{DS,1}, \ldots, \delta V_{DS,M}]^{\mathsf{T}}$ and $\delta \mathbf{V}_T = [\delta V_{T,1}, \ldots, \delta V_{T,M}]^{\mathsf{T}}$. The matrices $\mathbf{A}$ and $\mathbf{B}$ are given by:

$$\mathbf{A} = \begin{bmatrix} 1 + r_1 \cdot g_{DS,1} & 1 & \cdots & 1 \\ 1 - r_2 \cdot g_{m,2} & 1 + r_2 \cdot g_{DS,2} & \cdots & 1 \\ \vdots & & & \vdots \\ 1 - r_M \cdot g_{m,M} & 1 - r_M \cdot g_{m,M} & \cdots & 1 + r_M \cdot g_{DS,M} \end{bmatrix}$$

$$\mathbf{B} = \begin{bmatrix} -r_1 \cdot g_{T,1} & 0 & \cdots & 0 \\ 0 & -r_2 \cdot g_{T,2} & \cdots & 0 \\ \vdots & & & \vdots \\ 0 & 0 & \cdots & -r_M \cdot g_{T,M} \end{bmatrix}$$

We used the following notations: $g_{DS,i} = \partial f_i/\partial V_{DS,i}, g_{m,i} = \partial f_i/\partial V_{GS,i}, g_{T,i} = \partial f_i/\partial V_{T,i}$ and $r_i = R_M \| g_{m,i}^{-1}$ (parallel resistance connection). All the components of the matrices $\mathbf{A}$ and $\mathbf{B}$ are considered at the operating point. Finally, substituting Eq. 4.10 in Eq. 4.8 results in the following expression, with $\mathbf{v} = [1, \ldots, 1]^{\mathsf{T}}$:

$$\delta I = -\frac{1}{R_M} \cdot \mathbf{v}^{\mathsf{T}} \cdot \mathbf{A}^{-1} \cdot \mathbf{B} \cdot \delta \mathbf{V}_T \tag{4.11}$$

The previous steps show how the product $\mathbf{A}^{-1} \cdot \mathbf{B}$ is derived for the considered circuit. Another way consists in using the nodal admittance matrix of the circuit as presented in [5]. The perturbative approach approximates δI as a linear combination of all $\delta V_{T,i}$'s. This approach will be completed in the following section by a stochastic component: assuming that we have a model for the stochastic distribution of $\delta V_{T,i}$'s in the defect-free and the corrupted cases, then we can derive the distributions of δI (Eq. 4.11) and I (Eq. 4.3) for the useful and noisy signals.

4.4 Stochastic Current Model

In this section we model the sensed current as the sum of three components, which are analyzed separately. For each current component, a stochastic model is derived. Then, the test requirements are expressed in terms of the established stochastic distributions.

4.4.1 Components of the Sensed Signal

We divide the sensed current into a useful and a noisy part. The useful signal (I_u) is the current that flows through a nanowire when the code corresponding to its pattern is applied. On the other hand, the noise can be generated by two different processes: intrinsically (I_i), or defect-induced (I_d).

When a nanowire is not activated by the applied code, then the transistors laying in its decoder part generate a sub-threshold current, which we call *intrinsic noise* of a single nanowire $I_{i,0}$. This current is normally several orders of magnitude smaller than the useful signal since $I_u/I_{i,0} \sim I_{on}/I_{off}$, where I_{on} and I_{off} represent the on- and off-currents of the SiNWFETs that form the decoder. However, since the array has a large number of nanowires producing an intrinsic noise, the total intrinsic noise can be significant compared to the useful signal: $I_i = N_{off} \times I_{i,0}$, where N_{off} is the number of non-activated nanowires.

We consider now the case of a nanowire that should be off when no variability is taken into account. The high variability makes some digits of the nanowire pattern flip up and down. Then, the pattern changes and the nanowire may become activated by the code corresponding to another nanowire. In this case, the nanowire with the valid pattern is called a *victim*, and the nanowires that undergo this type of defects produce a *defect-induced noise*. Their number is denoted by N_{def}. Since the total number of nanowires is N, the following equation must hold $N_{use} + N_{off} + N_{def} = N$, where $N_{use} = 0$ if no nanowire is activated by the applied code, and $N_{use} = 1$ otherwise.

4.4.2 Distribution of the Useful Signal

We used the same model for V_T variation as in Sect. 3.3.4: every V_T is considered as an independent and normally distributed stochastic variable with mean value $\overline{V}_T$ and standard deviation σ_T. For instance, if a binary code is used, then $\overline{V}_T = V_{T,Ref0}$ holds for digits with the value 0, and $\overline{V}_T = V_{T,Ref1}$ holds for digits with the value 1. The distribution can be noted in the following way: $V_T \sim \mathcal{N}(\overline{V}_T, \sigma_T^2)$, and the same notation holds for the other (one- or multi-dimensional) stochastic variables throughout this chapter. In order to simplify the notations, V_T is discretized (for instance 0 and 1, assuming that V_T takes two values). If the nanowire pattern is correct, then the operating point of V_T coincides with its mean value. However, if a defect happens so that the bit representing V_T flips (for instance 1 becomes 0), then the operating point of V_T is shifted from the mean value of V_T by $-\Delta V_T$.

We consider a nanowire with a defect-free pattern **a**, which is controlled by its corresponding code $\mathbf{c}^a$, and which generates the useful signal I_u. Because the pattern is defect free, the operating points for the $V_{T,i}$'s are their respective mean values: $\mathbf{V}_T^{OP} = \overline{\mathbf{V}}_T$. Given that $\delta\mathbf{V}_T = \mathbf{V}_T - \mathbf{V}_T^{OP}$, then $\delta\mathbf{V}_T \sim \mathcal{N}(\mathbf{0}, \sigma_T^2 \cdot \mathbf{v})$. A useful signal flows through a nanowire with a given pattern on which a code corresponding to this pattern is applied. Thus, a useful signal follows the distribution resulting from Eqs. 4.3 and 4.11. The operating point is the on-current of the transistors I_{on}, which is calculated with SPICE simulator; whereas the variable part is given by Eq. 4.11 by applying the summation rule of independent and normally distributed variables. From the distribution of $\delta \mathbf{V}_T$ established here, it follows that:

$$\begin{cases} \delta I_u \sim \mathcal{N} \left(\bar{I}_{\delta u}, \sigma_{\delta u}^2 \right) \\ \bar{I}_{\delta u} = 0 \\ \sigma_{\delta u} = \dfrac{\sigma_T}{R_M} \cdot \| \mathbf{v}^\mathsf{T} \mathbf{A}^{-1} \mathbf{B} \| \end{cases} \tag{4.12}$$

Since $I_u = I_u^{OP} + \delta I_u, I_u$ can be modeled as a normal variable following the distribution f_u with the following parameters:

$$\begin{cases} I_u \sim \mathcal{N} \left(\bar{I}_u, \sigma_u^2 \right) \\ \bar{I}_u = I_{on} \\ \sigma_u = \dfrac{\sigma_T}{R_M} \cdot \| \mathbf{v}^\mathsf{T} \mathbf{A}^{-1} \mathbf{B} \| \end{cases} \tag{4.13}$$

We notice that the tail of the distribution of I_u for $I_u < 0$ has no physical meaning, because I_u cannot be negative. However, for reasonable and realistic values of σ_T and I_{on}, the probability that I_u, as calculated with this model, falls within this forbidden region is negligible.

4.4.3 Distribution of the Defect-Induced Noise

Now we consider a nanowire NW^b with the pattern $\mathbf{b}$ that undergoes some defects and turns into $\mathbf{b}^*$. This defective nanowire can be activated by the code $\mathbf{c}^a$ of another nanowire NW^a having the pattern $\mathbf{a}$. In this case, the defective nanowire generates a defect-induced noise I_d. Let n_0 be the number of digits 0 in $\mathbf{a}$ and S the set of indexes of digits with the value 0 in $\mathbf{a}$ that has to be turned into 1 in order to get $\mathbf{b}$: $S = \{i_1, \ldots, i_\tau\}$. Every transistor in the decoder part of the nanowire NW^b corresponding to one of these τ digits blocks the current flow through itself when $\mathbf{c}^a$ is applied on it in the defect-free case. In the case of defects, $\mathbf{b}$ turns to $\mathbf{b}^*$. In order to make NW^b with the defective pattern $\mathbf{b}^*$ conduct under application of $\mathbf{c}^a$, the blocking transistors need to have their V_T shifted down by ΔV_T; i.e. the digits of $\mathbf{b}$ considered above need to flip to 0.

The operating point of a transistor i having its $V_{T,i}$ shifted by ΔV_T is $\bar{V}_{T,i} - \Delta V_T$. For this transistor holds: $\delta V_{T,i} \sim \mathcal{N}(\Delta V_T, \sigma_T^2)(i \in S)$. For other transistors holds: $\delta V_{T,i} \sim \mathcal{N}(0, \sigma_T^2)(i \in \{1, \ldots, M\} \setminus S)$. We describe the series of shifts at all digits by the vector $\mathbf{s} \in \{0, 1\}^M$, where $\Delta V_T \cdot s_i \in \{0, \Delta V_T\}$ indicates whether a threshold voltage shift happened at the transistor i ($i = 1 \ldots M$). Then, by applying the summation rule of independent stochastic variables on Eq. 4.11, we get the distribution of the small signal of the defect-induced noise generated by one single nanowire characterized by a given vector $\mathbf{s}_1$:

$$\begin{cases} \delta I_{d,1} \sim \mathcal{N} \left(\bar{I}_{\delta d,1}, \sigma_{\delta d,1}^2 \right) \\ \bar{I}_{\delta d,1} = -\dfrac{\Delta V_T}{R_M} \cdot \mathbf{v}^\mathsf{T} \mathbf{A}^{-1} \mathbf{B} \cdot \mathbf{s}_1 \\ \sigma_{\delta d,1} = \dfrac{\sigma_T}{R_M} \cdot \| \mathbf{v}^\mathsf{T} \mathbf{A}^{-1} \mathbf{B} \| \end{cases} \tag{4.14}$$

Here again, the operating point for $I_{d,1}$ is the same as before, I_{on}, since all the transistors of the decoder part of this defective nanowire are switched on. Consequently, the distribution of $I_{d,1}$ can be expressed as follows:

$$\begin{cases} I_{d,1} \sim \mathcal{N} \left(\bar{I}_{d,1}, \sigma_{d,1}^2 \right) \\ \bar{I}_{d,1} = I_{on} - \dfrac{\Delta V_T}{R_M} \cdot \mathbf{v}^\mathsf{T} \mathbf{A}^{-1} \mathbf{B} \cdot \mathbf{s}_1 \\ \sigma_{d,1} = \dfrac{\sigma_T}{R_M} \cdot \| \mathbf{v}^\mathsf{T} \mathbf{A}^{-1} \mathbf{B} \| \end{cases} \tag{4.15}$$

The number of nanowires that generate the defect-induced noise depends on the variability level of the technology. If we assume that N_{def} nanowires generate a defect-induced noise, then every one of them is characterized by a given threshold voltage shift vector $\mathbf{s}_i, i \in \{1, \ldots, N_{def}\}$ and a distribution:

$$\begin{cases} I_{d,i} \sim \mathcal{N} \left(\bar{I}_{d,i}, \sigma_{d,i}^2 \right) \\ \bar{I}_{d,i} = I_{on} - \dfrac{\Delta V_T}{R_M} \cdot \mathbf{v}^\mathsf{T} \mathbf{A}^{-1} \mathbf{B} \cdot \mathbf{s}_i \\ \sigma_{d,i} = \dfrac{\sigma_T}{R_M} \cdot \| \mathbf{v}^\mathsf{T} \mathbf{A}^{-1} \mathbf{B} \| \end{cases} \tag{4.16}$$

Since the total defect-induced noise is equal to the sum of all defect-induced noises generated by single nanowires, the distribution of the total defect-induced noise generated by N_{def} nanowires follows the normal distribution $f_d^{N_{def}}$ given by the following parameters:

$$\begin{cases} I_d^{N_{def}} \sim \mathcal{N} \left(\bar{I}_d^{N_{def}}, (\sigma_d^{N_{def}})^2 \right) \\ \bar{I}_d^{N_{def}} = N_{def} \cdot I_{on} - \dfrac{\Delta V_T}{R_M} \cdot \mathbf{v}^\mathsf{T} \mathbf{A}^{-1} \mathbf{B} \cdot \displaystyle\sum_{i=1 \ldots N_{def}} \mathbf{s}_i \\ \sigma_d^{N_{def}} = \dfrac{\sqrt{N_{def}} \cdot \sigma_T}{R_M} \cdot \| \mathbf{v}^\mathsf{T} \mathbf{A}^{-1} \mathbf{B} \| \end{cases} \tag{4.17}$$

4.4.4 Distribution of the Intrinsic Noise

The intrinsic noise is generated in the subthreshold regime of the transistors forming the decoder part of the nanowire. If a nanowire that is supposed to be switched off, partially or totally turns on because of defects affecting his pattern, then it is considered to be generating a defect-induced noise. In contrast, if the defects shift the threshold voltages of the nanowire to higher values and make it more resistive, then the nanowire generates less intrinsic noise. In the worst case, the maximum intrinsic noise of a single nanowire $I_{i,0}$ is equal to the off-current of the transistors fabricated with the considered technology (I_{off}). As explained at the beginning of this section, the total intrinsic noise is equal to the sum of all signals generated by the N_{off} transistors that are switched off. Thus, $I_i = N_{off} \times I_{off}$ is the maximum expected intrinsic noise, which is the worst-case consideration of the intrinsic noise, as a constant additive current, utilized in this work.

4.4.5 Model of the Test Requirements

Given the electrical expressions and the stochastic distributions of the different components of the sensed signal, it is possible now to express analytically the probabilities P_0, P_1 and P_2 (Sect. 4.2.2) in order to estimate the error probability of the test depending on the technology and the circuit parameters.

If no nanowire is addressed, then $N_{use} = 0, N_{off} = N, N_{def} = 0$ and $I_s = N \cdot I_{off}$. The probability P_0 of detecting that no nanowire is addressed, is simply equal to 1 if I_0 is set greater than $N \cdot I_{off}$, otherwise it is equal to 0. Thus, P_0 can be expressed as follows, where $\delta(x)$ represents the Dirac distribution around 0:

$$P_0 = \int_{-\infty}^{I_0} \delta(I - N \cdot I_{off}) \, dI \tag{4.18}$$

If we consider the case that one single nanowire is addressed, then $N_{use} = 1, N_{off} = N - 1, N_{def} = 0$ and $I_s = I_u + (N - 1) \cdot I_{off}$. The additional term $(N - 1) \cdot I_{off}$ shifts the mean value of I_u by $(N - 1) \cdot I_{off}$; or equivalently, it shifts the borders of the integral P_1 by $(N - 1) \cdot I_{off}$. Thus, P_1 can be expressed as follows:

$$P_1 = \int_{I_0 - (N-1) \cdot I_{off}}^{I_1 - (N-1) \cdot I_{off}} f_u(I) \, dI \tag{4.19}$$

Now, we consider the last case in which one or more nanowires are generating defect-induced noise. Here $N_{use} = 1$ and $N_{off} = N - 1 - N_{def}$ hold; where N_{def} depends on the variability level of the technology. The sensed signal is given by: $I_s = I_u + I_d + (N - 1 - N_{def}) \cdot I_{off}$. Let B_i be the event that exactly i nanowires are generating a defect-induced noise, $B = \cup B_i$ and $A = \{I_1 \leq I_s\}$. Then, by using the Bayesian relations, we obtain:

$$
\begin{aligned}
P_2 &= \Pr\{A|B\} \\
&= \Pr\frac{\{A \cap B\}}{\Pr\{B\}} \\
&= \frac{1}{\Pr\{B\}} \times \sum_{i=1...N_{\text{def}}} \Pr\{A|B_i\} \cdot \Pr\{B_i\}
\end{aligned}
\tag{4.20}
$$

The expression $\Pr\{A|B_i\}$ represents the conditional probability that we detect the defect-induced noise generated by one or more nanowires, given the fact that there are exactly i nanowires generating this kind of noise, with $i = 1 \ldots N_{\text{def}}$. Equation sets 4.13 and 4.17 give the distributions that can be used to calculate $\Pr\{A|B_i\}$. Because of the intrinsic noise, the borders of the integral P_2 are shifted by $(N - 1 - N_{\text{def}}) \cdot I_{\text{off}}$:

$$
\Pr\{A|B_i\} = \int_{I_1-(N-1-i)\cdot I_{\text{off}}}^{+\infty} (f_{\text{u}} + f_{\text{d}}^i)(I)\, \mathrm{d}I
\tag{4.21}
$$

The symbol $(f_{\text{u}} + f_{\text{d}}^i)$ denotes the distribution resulting from the sum of both independent and normal distributions f_{u} and f_{d}^i, which is also a normal distribution with the mean value $\bar{I}_{\text{u}} + \bar{I}_{\text{d}}^i$ and the standard deviation $\sqrt{\sigma_{\text{u}}^2 + \sigma_{\text{d}}^{i2}}$. In order to calculate $\Pr\{B_i\}$, we refer to Algorithm 3 in Chap. 3, which enumerates all possible defect scenarios for a given variability level and calculates their respective probabilities. This algorithm was used for hot code, and the construction rule can be relaxed in order to address the case of tree and reflexive code. The obtained results are used in order to estimate $\Pr\{B_i\}$ and $\Pr\{B\} = \sum \Pr\{B_i\}$.

4.5 Model Implementation

We implemented the circuit by using the bulk MOSFET model for the considered SiNWFET, as described in [6]. The linearization around the operating point was performed in the linear region, in order to keep $V_{\text{DS},i}$, and consequently V_{P}, as low as possible.

It is desirable to obtain a symmetrical device operation, i.e., the same value of the operating point at all transistors, in order to simplify the matrices **A** and **B**. In order to make the operation symmetrical, it is necessary to set some conditions on the device technology and the electrical bias of the circuit. We explained in Sect. 4.3 that the technology used modifies more the threshold voltages than the expression of the output function f of the transistors, i.e., the gain factors β_i can be considered to be identical for all transistors. On the other hand, all $V_{\text{DS},i}$ have the same value at the operating point, when all transistors are biased in the same way, i.e., when all

$V_{GS,i} - V_{T,i}$ at the operating point are equal. Thus, we consider only circuits biased in this symmetrical way, and then, we can omit the index $i = 1 \cdots M$ in matrices **A** and **B**.

The condition that insures that the devices are biased in the linear region is given for short-channel transistors by: $V_{DS} < (1 - \kappa)(V_{GS} - V_T)$, with $\kappa = 1/(1 + E_{sat}L/(V_{GS} - V_T))$; E_{sat} being the electrical field at saturation and L the channel length. We verified this condition during the simulations of the operating points. We also noticed that V_{DS} was small enough to insure that $R_M \cdot g_m \ll 1$ for the considered range of realistic values of R_M and β. Then, $R_M \| g_m^{-1} \approx R_M$ holds.

Given the fact that the current expression described by f depends on $V_{GS} - V_T$ but not V_{GS} nor V_T alone, then it can be assumed with a very good accuracy that $-g_T = g_m$, which is the known transconductance of the transistors. In this case, **A** and **B** can be simplified to:

$$\mathbf{A} = \mathbf{v} \cdot \mathbf{v}^\mathsf{T} + R_M \cdot g_{DS} \cdot \mathbf{I}$$
$$\mathbf{B} = R_M \cdot g_m \cdot \mathbf{I}$$

with **I** the $M \times M$ identity matrix, and **v** the previously introduced vector with M entries equal to 1. Then, the expression of **A** can be easily inverted by using the binomial inverse theorem: $(\mathbf{v} \cdot \mathbf{v}^\mathsf{T} + k \cdot \mathbf{I})^{-1} = k^{-1} \cdot \mathbf{I} - (k^2 + k \cdot M)^{-1} \cdot \mathbf{v} \cdot \mathbf{v}^\mathsf{T}$. This yields the following expressions:

$$\mathbf{A}^{-1} = \frac{1}{R_M \cdot g_{DS}} \cdot \mathbf{I} - \frac{1}{R_M \cdot g_{DS} \cdot (R_M \cdot g_{DS} + M)} \cdot \mathbf{v} \cdot \mathbf{v}^\mathsf{T}$$

And the ultimate product, which is needed in all current distributions is given by:

$$\mathbf{v}^\mathsf{T} \mathbf{A}^{-1} \mathbf{B} = \frac{R_M \cdot g_m}{R_M \cdot g_{DS} + M} \mathbf{v}^\mathsf{T} \tag{4.22}$$

While the distribution of the useful signal for the considered practical implementation of the circuit is fully described by inserting the expression above into Eq. 4.13, the distribution of the defect-induced noise still needs other parameters in order to be fully described: the V_T-shift vectors $\mathbf{s}_i$ and the number of defective and noisy nanowires N_{def}. These parameters are analyzed next.

In the implemented circuit, we considered a binary reflexive code with the length M. Such a code has M digits in $\{0, 1\}$ and two complementary halves with the length $M/2$ each, i.e., if the digit $c_i = 0$ then $c_{M/2+i} = 1$ and vice versa. We consider now a vector **s** representing the positions of the τ shifts of V_T, which make a defective pattern noisy, i.e., activated by the code of a victim pattern. The number of shifts τ can be estimated in average. An eligible V_T-shift vector **s** gives the positions of the τ digits 1 in the noisy pattern that turned to 0. Because of the symmetry of both halves of the reflexive code, τ can be any number in $1 \dots M/2$.
For every value of τ there are $\begin{pmatrix} M/2 \\ \tau \end{pmatrix}$ possible noisy pattern. The average value of τ is thus given by the binomial theorem: $\bar{\tau} \approx M/4$.

Now, we can calculate the following product:

$$\mathbf{v}^\mathsf{T}\mathbf{A}^{-1}\mathbf{B} \cdot \mathbf{s}_i = \frac{R_\mathrm{M} \cdot g_\mathrm{m} \cdot \tau_i}{R_\mathrm{M} \cdot g_\mathrm{DS} + M}$$

where τ_i is the number of V_T-shifts described by the vector $\mathbf{s}_i$. Consequently, the following result based on the average value of τ_i holds:

$$\mathbf{v}^\mathsf{T}\mathbf{A}^{-1}\mathbf{B} \sum_{i=1}^{N_\mathrm{def}} \mathbf{s}_i \approx \frac{R_\mathrm{M} \cdot g_\mathrm{m} \cdot N_\mathrm{def} \cdot \bar{\tau}}{R_\mathrm{M} \cdot g_\mathrm{DS} + M} \tag{4.23}$$

By inserting this result into Eq. 4.17, we obtain the distribution of the noise generated by N_def defective nanowires in the implemented circuit.

At this level, both P_0 and P_1 can be calculated for the implemented circuit. However, in order to calculate P_2, we need to estimate the value of Eq. 4.21 for all possible values of the number of defective nanowires. This is performed the following way. There are potentially N defective nanowires, and their average number can be estimated with Algorithm 3 in Chap. 3 used exhaustively for large number of covers (≥ 1) of every defective code, and relaxed with respect to the code construction rule in order to include reflexive codes in addition to hot codes. Since the contribution of the probability given by Eq. 4.21 is weighted by $\Pr\{B_i\}$ (the probability that exactly i nanowires are noisy), we estimated with Algorithm 3 these weights under different conditions of the variability (σ_T). Our experimental results show that $\Pr\{B_1\}$ is several orders of magnitudes larger than $\Pr\{B_i\} \,\forall\, i \geq 2$ for realistic values of σ_T. Consequently, we can consider solely the contribution of $\Pr\{B_1\}$ to P_2 (i.e., $N_\mathrm{def} = 1$) while keeping the model accurate enough.

Finally, the useful signal in the implemented circuit follows the distribution f_u defined as follows:

$$\begin{cases} I_\mathrm{u} \sim \mathcal{N}\left(\bar{I}_\mathrm{u},\ \sigma_\mathrm{u}^2\right) \\[2mm] \bar{I}_\mathrm{u} = I_\mathrm{on} \\[2mm] \sigma_\mathrm{u} = \dfrac{\sigma_\mathrm{T} \cdot g_\mathrm{m} \cdot \sqrt{M}}{R_\mathrm{M} \cdot g_\mathrm{DS} + M} \end{cases} \tag{4.24}$$

The defect-induced noise in the implemented circuit follows the distribution f_d defined by the following parameters:

$$\begin{cases} I_\mathrm{d} \sim \mathcal{N}\left(\bar{I}_\mathrm{d},\ \sigma_\mathrm{d}^2\right) \\[2mm] \bar{I}_\mathrm{d} = I_\mathrm{on} - \dfrac{\Delta V_\mathrm{T} \cdot g_\mathrm{m} \cdot M/4}{R_\mathrm{M} \cdot g_\mathrm{DS} + M} \\[2mm] \sigma_\mathrm{d} = \dfrac{\sigma_\mathrm{T} \cdot g_\mathrm{m} \cdot \sqrt{M}}{R_\mathrm{M} \cdot g_\mathrm{DS} + M} \end{cases} \tag{4.25}$$

The probabilities P_0 and P_1 used to estimate the test error remain unchanged, whereas P_2 is simplified as follows (by using the same conventional notation defined in Sect. 4.4.5 for $(f_u + f_d)$, the sum of the normal distributions f_u and f_d):

$$
\begin{cases}
P_0 = \displaystyle\int_{-\infty}^{I_0} \delta(I - N \cdot I_{\text{off}})\, dI \\[2ex]
P_1 = \displaystyle\int_{I_0-(N-1)\cdot I_{\text{off}}}^{I_1-(N-1)\cdot I_{\text{off}}} f_u(I)\, dI \\[2ex]
P_2 = \displaystyle\int_{I_1-(N-2)\cdot I_{\text{off}}}^{+\infty} ((f_u + f_d)(I))\, dI
\end{cases}
\tag{4.26}
$$

4.6 Simulation Results

In this section we first investigate the statistical behavior of the sensed signal. Then, we show the validation of the previously established analytical results to estimate the optimal values of the thresholder parameters I_0 and I_1. Next, we present simulations to evaluate the test quality ϵ, defined as the test error probability. Finally, as the proposed testing approach relies on the linearization of a circuit around an operating point, we perform a complete exploration of the incurred linearization error.

In the simulations we have fixed the supply voltage to $V_P = 0.9\,\text{V}$ in order to fulfil the assumption that the transistors are in the linear region and we checked this value with simulations. We noticed that increasing V_P beyond 1 V causes V_{DS} at the operating point to increase and the admissible range for linear operation of the transistors to shrink. Since we used a binary code, only two threshold voltages were needed. Then, ΔV_T was set to V_P and $V_{\text{GS}} - V_T$ to $\Delta V_T/2$ for a maximum voltage-bias at the gate. We assumed that the mesowires are defined on the 65-nm technology node, which is half pitch of the poly-silicon gates as drawn in the memory design [7]. Then, the technology parameter β was not fixed. Its value depends on the dimensions and quality of the nanowires, which can be as strong as $100\,\mu\text{AV}^{-2}$ or as weak as $10\,\mu\text{AV}^{-2}$. In fact, it is likely to obtain weak SiNWFETs when narrow nanowires are used because of the small shape ratio W/L ($W \ll L$, with W and L the channel width and length of the SiNWFET respectively). The additional technology parameters that were explored during the simulations are the resistance of the nanowire in the memory part R_M and the variability level expressed as the standard deviation of the SiNWFET threshold voltage σ_T. The number of mesowires M was also left as a design parameters, because it depends on the size of the memory.

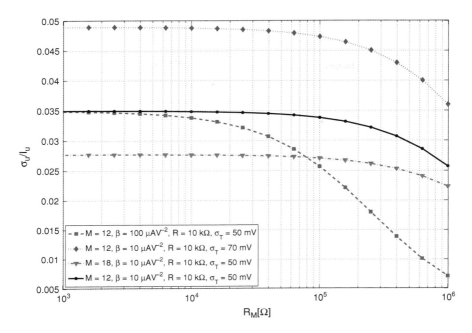

Fig. 4.6 $\sigma_u/\bar{I}_u$ vs. design and technology

4.6.1 General Signal Variation

We investigated the statistical behavior of the sensed signal, first by simulating the ratio $\sigma_u/\bar{I}_u$ under different conditions. The results are shown in Fig. 4.6. Increasing the resistance always improves this ratio, because σ_u decreases with decreasing R_M faster than $\bar{I}_u$ does. Increasing β also improved this ratio, because a higher current injection occurs. As expected, this ratio degrades when σ_T increases, because it scales linearly with σ_T. Increasing M is shown to be beneficial, because for reasonable values of the operating point $R_M \cdot g_{DS} \ll M$, thus $\sigma_u \sim 1/\sqrt{M}$. The defect induced noise was investigated in the same way, and we noticed a similar qualitative behavior of $\sigma_d/\bar{I}_d$, which was about twice higher than $\sigma_u/\bar{I}_u$.

4.6.2 Optimization of the Thresholder Parameters

The thresholder parameters that we are investigating in this work are I_0 and I_1. The minimal value of I_0 is given by P_0 in Eq. 4.26: I_0 has to be greater than $N \cdot I_{off}$ in order to insure that $P_0 = 1$. While keeping I_0 larger than this critical value, we plotted I_1 that gives the best test quality (i.e., the minimal error ϵ). The results are shown in Fig. 4.7 for different technology and design parameters. The staircase

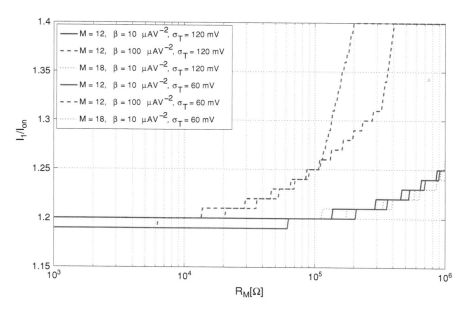

Fig. 4.7 Optimal value of I_1 vs. design and technology

shape of the plot is due to the reverse numerical calculation of the integral borders in Eq. 4.26. I_1/I_{on} increases with R_M and β; which can be explained by the fact that the distributions of the noise and useful signal become more centred around their respective mean values as shown in the previous simulation. Since the noise has a wider distribution, it has the strongest impact on I_1 when both useful and noisy current distributions become thinner; then it moves I_1 towards $\bar{I}_d$ by increasing it slightly. Among the considered technology parameters, β has the strongest influence on I_1. However this influence is globally weak: for $R_M = 10\,k\Omega$, increasing β by a factor of 10, adds just 4% to I_1. Larger values of R_M cause I_1 to increase by more than 4%. Though, it is unlikely to have both β and R_M large; because β increases with the nanowire width W; while the opposite happens to R_M. If we fix all the technological parameters (β, R_M and σ_T); then increasing the design parameter M from 12 to 18 has less impact than increasing β by a factor of $10\times$, because $\sigma \sim \beta/\sqrt{M}$, showing that the dependency on M is weaker. Consequently, I_1 has a robust value $\sim 1.2 \times I_{on}$ that depends on the technology by less than 4%.

Once I_1 was calculated, we fixed the value of I_0 to $I_1/(1 + \epsilon)$ for a given positive ϵ. We noticed that for a small ϵ around 0.5 or less, the test quality degraded, because the thresholder range is too narrow to separate noise from useful signal. For ϵ larger than 0.7, the test quality remains constant because the thresholder range is large enough. But if it becomes too large, then the intrinsic noise cannot be separated from the useful signal anymore and the test quality degrades again. Thus, I_0 should be large enough compared to the intrinsic noise I_i, which implies $\epsilon \ll (1.2/N \cdot I_{on}/_{off} - 1)$. For a wide range of reasonable technological assumptions and memory size, ϵ can be set to 0.8, i.e., $I_0 \sim 0.66 \times I_{on}$.

4.6.3 Analysis of Test Quality

By using these optimized thresholder parameters, we investigated the test quality under different conditions. The test quality is defined as the minimum test error, as plotted in Fig. 4.8. As expected, the best test quality is obtained for $I_1 \sim 1.2 \times I_{on}$; the position of the minimum is almost insensitive to the technology. A typical set of parameters, as expected for this technology, is $\beta = 10\,\mu A\,V^{-2}$, $R = 10\,k\Omega$ and $\sigma_T = 70$ mV. For a small-granularity array with $M = 12$, the test error is $\epsilon \sim 10^{-4}$. Reducing the power level from 0.9 V down to 0.6 V reduces the current level at the operating point without reducing its variable part. Thus, it increases the noise level in the sensed current, and the test quality degrades by a factor of 22×. Consequently, the power level should be kept as high as possible under the test conditions. The variability level is the most critical parameter: increasing σ_T to 100 mV degrades the test quality by a factor larger than 50×. Improving the transistor gain factor β by 10× enhances the test quality by a factor of 3×. A 20× higher memory resistance R_M improves the test quality by a factor of 10×.

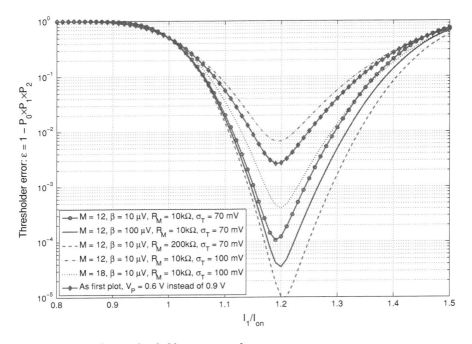

Fig. 4.8 Test quality vs. thresholder parameter I_1

However, this is not a useful strategy, because the memory should be designed with the lowest possible R_M, in order to obtain a higher level for the sensed current and to facilitate the read-out procedure. In fact, our analytical model and results show that a better strategy is to increase the number of addressing wires M: by using 50% more addressing wires, the test quality improves by 18× under the same technological assumptions.

The design parameter M plays a major role not only in addressing the array, but also in improving the test quality. Increasing the number of mesowires M without changing the number of addressed nanowires adds redundancy to the decoder circuit and it guarantees an improvement of the test quality by reducing the test error. Such an increase of the number of mesowires signifies a choice of a redundant code, which is longer than the minimal code needed to address the N available nanowires. We considered the same array of nanowires having codes with different lengths; and we estimated the area overhead with the model explained in Chap. 3 and the test quality as shown in the previous figure. The results are summarized in Table 4.1. For instance, adding 6 more digits to the initial 12 digits represented by 12 mesowires, gives a longer code, i.e., an area overhead of 35.5% and an improvement of the test quality by a factor of 59.2× for $\sigma_T = 80$ mV and 15.2× for a higher variability ($\sigma_T = 100$ mV). It is therefore important to consider the expected test quality, while designing the decoder of the memory array.

4.6.4 Exploration of Linearization Error

In the final set of experiments we have explored how high the linearization error is. In order to keep the assumption of having a circuit in the neighborhood of a symmetric operating point accurate enough, we should insure that the deviation from this operating point is not too large. We quantified the variation of V_{DS}'s from their operating points depending on the variation of V_T's from their operating points. The choice of V_{DS}'s was motivated by the fact that the system presented in Sect. 4.3 was solved for the state vector $\delta\mathbf{V}_{DS}$, which yields the current variation δI. The sensitivity of V_{DS} to V_T is defined as $(\delta V_{DS}/V_{DS}) \times (\delta V_T/V_T)$, which is almost equal to $\eta = \partial \log V_{DS}/\partial \log V_T$. Figure 4.9 depicts the variation of η depending on the design and technology parameters. For instance, $\eta = 0.2$ means that when V_T varies by $x\%$, then V_{DS} varies by $0.2x\%$. The threshold voltage

Table 4.1 Area/test quality trade-off: area overhead and test quality improvement by adding redundancy. Notice that $M = 12$ is the reference without redundancy

M	12	14	16	18
Area overhead (%)	0	11.8	23.6	35.5
Improv. of ϵ				
$\sigma_T = 80$ mV	1×	3.9×	14.5×	59.2×
$\sigma_T = 100$ mV	1×	2.5×	6.1×	15.2×

variation δV_T has the same order of magnitude as σ_T. Assuming that the process is unreliable and $\sigma_T/V_T = 20\%$, then δV_{DS} is just 4% of V_{DS} in the worst case (R_M very large) of the typical scenario ($\beta = 10\,\mu A\,V^{-2}$ and $M = 12$). Figure 4.9 shows that the model is precise enough for either strong and weak SiNWFETs and for both memory sizes considered here ($M = 12$ and $M = 16$). There is no amplification of the variability (i.e., $|\eta| < 1$); however the model loses some accuracy when both R_M and β are large. As explained before, from the physical point of view, R_M changes opposite to β when the nanowire geometry varies. Generally, the technology is optimized in order to obtain the smallest possible R_M.

The linearization error causes the transistors to have different operating points and breaks the assumed symmetry of the circuit: the assumption $V_{DS,i}$ are all equal at the operating point $\forall i$, does not hold anymore. The slight shift of the operating point causes the product $\mathbf{v}^T\mathbf{A}^{-1}\mathbf{B}$ to deviate from the calculated value. Thus, the error propagates to σ_u and σ_d; and finally it induces a different test quality. Interestingly, the optimal test is still obtained for $I_1 = 1.2 \times I_{on}$ as explained above: Unlike the test quality, the design of the thresholder is not affected by the linearization error. In Table 4.2 we summarized the reduction of the estimated test quality due to the linearization error for different scenarios. The strongest reduction of the test quality (by a factor of $0.47\times$) occurs for high β. In all other scenarios, the linearization error is less important, and the calculated test quality is almost equal to the actual value. In all these cases, the reduction of the test quality

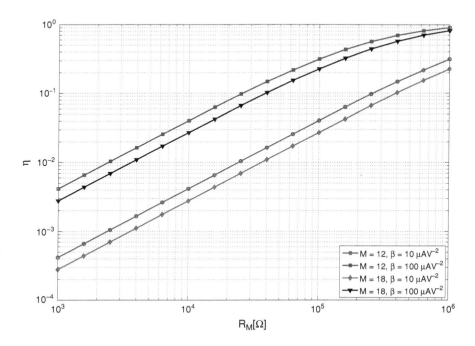

Fig. 4.9 Sensitivity η vs. design and technology

Table 4.2 Reduction of the calculated test quality caused by the linearization error

Scenario	M	$\beta\,(\mu AV^{-2})$	σ_T (mV)	$\epsilon^{calc}/\epsilon^{act}$
S1	12	10	70	0.85×
S2	12	50	70	0.47×
S3	12	10	100	0.93×
S4	18	10	70	0.80×

due to the linearization error is 5 to 60× smaller than the improvement of the test quality by choosing a redundant design, as shown in Table 4.1.

4.7 Discussions

As already mentioned in the previous chapter, the physical defects are mainly based on the dopant variation in small nanowire dimensions. The model can be enhanced by including other defects such as the nanowire cross-section and the oxide thickness variation, which defects have an impact on the current flowing through the nanowire. Such variations can be included in the model by extending the variational vector $[\delta V_{T,1}, \ldots, \delta V_{T,M}]^\top$ to other variable parameters, and by increasing the size of the matrices $\mathbf{A}$ and $\mathbf{B}$ accordingly. The advantage of this approach is that it gives a more accurate result; however, the symmetry of the matrices is lost, which results in non-explicit expressions for the distribution of the sensed current. Nevertheless, these expressions can be still evaluated using suitable computing tool even if they are not explicit.

There is another question related to the model, namely about the necessity of discovering defective nanowires addressed by the same codes. It may be interesting to simplify the test procedure and consider such nanowires as working in parallel and at the same time, i.e., they seem to be a single nanowire with larger dimensions (lower resistance, higher capacitance). However, this simplified approach implies that these *large* nanowires would control more than one bit. Some single nanowires can be shared between *large* nanowires, resulting in a mutual control of the addressed bits by nanowires not belonging to the same *large* nanowires. This results in turns in the PSF described before, whose detection may be time-consuming. It is therefore necessary to detect defective nanowires sharing the same address, as proposed in this model.

In the proposed model, many assumptions are introduced in order to get simplified expressions of the distribution of the different current components; and these assumptions are verified during the simulations. It is important to highlight the implication of such assumptions on the model and on the practical implementation of such test procedure. On the one hand, the symmetrical operation of all transistors requires that all applied voltages are shifted by the voltage drop between the source of the considered transistor and the ground. This drop is not easy to estimate accurately, and it necessitates the use of an additional level-shifter circuitry. Besides that, SiNWFET with floating bodies may make such control less accurate. On the other hand, the gain factors β_i are most likely slightly variable, depending on the doping level. A more accurate model has to include their variability as well in the matrices **A** and **B**. However, the overall methodology presented in this chapter remains valid.

The off-current I_{off} is an important model parameter that vanishes in the dimensioning rules for the thresholder, because it is assumed to be low enough. The overall off-current increases with the size of the crossbar and the variability level. Even if the crossbar is average in size, for instance having a few tens of nanowires in every layer, it is important to check the assumption on the level of I_{off}, by using an on-chip calibration crossbar that measures the typical value of the off-current in the crossbar. If the total off-current is not negligible, then it has to be included in the dimensioning rules, by shifting the values of I_0 and I_1 by the noise level generated by the off-current.

In this chapter, it is implicitly assumed that defects happen only in the crossbar. The CMOS part is assumed to be defect-free. This is not a very accurate assumption, since dopant variation is also affecting the CMOS part as well. But the crossbar variability is more noticeable given the fact that the nanowire thickness is assumed to be much smaller than the photolithography half-pitch. The variation of the CMOS part can be seen as a variation of the voltages applied to the decoder, i.e., the V_A's. This was modeled in Chap. 3, showing that the consequences can be very important if the overall variability is very large. It is therefore possible to extend the proposed decoder test model by including the variation of V_A caused by the CMOS part of the circuit.

4.8 Chapter Contributions and Summary

This chapter presents a novel approach to test nanowire decoders. The approach is motivated by the need to test the nanowire decoder separately before the whole system is tested in order to simplify the overall testing complexity. The decoder test procedure saves desirable addresses in a LUT, and its operating principle is based on the current level through the nanowire.

In this chapter, an analytical model for the sensed current is presented. It is based on the linearization of the current at the operating point, assuming that small variability represents a variation captured by the sensitivity model, while defects are modeled as a shift of the operating point. The model includes the useful current, the intrinsic noise due to off-transistors and the defect-induced noise generated by transistors operating at an undesirable operating point.

The implementation of the model gives the dimensioning rules for the thresholder used in the decoder test. These rules are robust with respect to the variation of the technology and design parameters, resulting in a defect-tolerant test operation. When the test is performed with the optimal thresholder parameters, then it results in low test error probabilities.

This chapter introduces an innovative way of performing design for test. As the test quality improves with the increasing level of redundancy in the decoder, it is possible to design the decoder in such a way that the expected test quality is maximized. Consequently, there are many opportunities to optimize the choice of the code type and length in order to improve the decoder test quality.

This chapter concludes the sequence of two chapters related to logic design of crossbar decoders, and it concludes the part of the book related to the nanowire crossbar technology. In the following chapter, a different technology based on carbon nanotubes is introduced, and the design of logic circuit is optimized by leveraging inherent properties of this technology.

References

1. Ben Jamaa MH, Atienza D, Leblebici Y, De Micheli G (2009) A stochastic perturbative approach to design a defect-aware thresholder in the sense amplifier of crossbar memories. In: Proceedings of ASP-DAC, pp 835–840
2. Abadir MS, Reghbati HK (1983) Functional testing of semiconductor random access memories. ACM Comput Surv 15(3):175–198
3. Luo Y, Collier CP, Jeppesen JO, Nielsen KA, DeIonno E, Ho G, Perkins J, Tseng H-R, Yamamoto T, Stoddart JF, Heath JR (2002) Two-dimensional molecular electronics circuits. J Chem Phys Phys Chem 3:519–525
4. Adams RD (2003) High performance memory testing. Kluwer Academic Publishers, Norwell
5. Brayton RK (1980) Sensitivity and optimization. Elsevier, Amsterdam
6. Rabaey JM (1996) Digital integrated circuits, a design perspective. Prentice Hall International Editions, Upper Saddle River
7. International technology roadmap for semiconductors (ITRS) (2006) Tech Rep. www.itrs.net/reports.html

Chapter 5
Logic Design with Ambipolar Devices

In this part of the book, the work is based on a different technology from the previous chapters. Carbon nanotubes are another candidate for the extension of CMOS by replacing the silicon channel. The CNT technology also promises novel ways to design logic circuits by leveraging its ability to on-line control the device polarity. Despite the expected large performance enhancement, there are still many issues related to this immature technology to be resolved.

Previous works have focused on CNTs as a bare replacement of silicon channel in MOSFETs. In other words, the mainstream research is dedicated to import MOSFET's design style and transfer it to CNTFETs. This approach is expected to help CMOS technology to progress two to three technology nodes forwards, in terms of performance (i.e., intrinsic delay). Most of the efforts have been targeting the improvement of the technology by better controlling the CNT growth and dispersion [1] and reducing the part of metallic CNTs within the grown structures [2]. Some of these challenges can be addressed along with the circuit design, by making the design more fault-tolerant [3–5].

Some of the previous approaches addressed the challenges and opportunities given by the fabrication of ambipolar CNTFETs [6, 7]. Such devices have two gates, one of them controls the current through the channel, while the other one controls the device polarity, i.e., whether it is n- or p-type. Given the additional on-line controllability of device polarity, novel logic design aspects that do not have any counterpart in CMOS technology have been considered as well [8, 9]. However, these works on the design level did not consider any possible logic operation that can involve both gate signals at the same time, but they rather considered the signal applied at the polarity control gate to be fixed or hard-wired.

This chapter focuses on logic design styles with ambipolar CNTFETs. The novelty of the approach is that it considers both gate signals as circuit-internal signals, on which independent logic operations can be performed. This opens up the opportunity to implement in a very efficient and compact way the XOR operation and many other complex logic functions embedding or not the XOR operation.

M. H. Ben Jamaa, *Regular Nanofabrics in Emerging Technologies*,
Lecture Notes in Electrical Engineering, 82, DOI: 10.1007/978-94-007-0650-7_5,
© Springer Science+Business Media B.V. 2011

This chapter leverages this novelty that does not exist in CMOS technology, or it may exist but for a higher price in terms of area and delay. The design of different dynamic and static logic families is proposed for the first time, and many variants of the static logic family are evaluated by comparing the results of multi-level logic synthesis on a benchmark of logic circuits. The other novelty of this part of the work is the design of a set of regular fabrics based on dynamic and static ambipolar CNTFET complex logic blocks, which can be used either for FPGA or regular ASIC design.

This chapter has been partly published in [10, 11] and it is organized as follows. First, previous work on fabrication and logic design with CNTFETs is surveyed. Then, the intrinsic properties of ambipolar CNTFETs are presented. In the following sections, both dynamic and static logic families with different variants are designed. Subsequently, the designed library of static gates and its variants are characterized and evaluated by synthesizing a benchmark of logic circuits and comparing its synthesis to the synthesis results with CMOS technology. After a discussion on the limits of the presented approach, the chapter is concluded with a summary of its contributions.

5.1 Logic Circuits with Carbon Nanotubes

CNTFETs are novel devices that are expected to sustain the transistor scalability while increasing its performance. One of the major differences between CNTFETs and MOSFETs is that the channel of the former devices is formed by CNTs instead of silicon, which enables a higher drive current density, due to the larger current carrier mobility in CNTs compared to bulk silicon and to the better gate control of the quasi one-dimensional CNTs [12]. Despite the similar structure of these two devices, CNTFETs necessitate the use of different gate and drain/source contact materials in order to optimize the threshold voltage and to reduce the Shottky barriers. It also necessitates different doping techniques [13].

CNTs suffer from a high variability with respect to their diameter and alignment. The manufacturing techniques also yield undistinguishable metallic and semiconducting CNTs. Several solutions have been proposed to address these problems, on both the manufacturing and the design level. It was demonstrated that a plasma oxygen etch has a higher selectivity to metallic CNTs compared to semiconducting ones [2]. In another work, the application of an electrical field during the CNT growth process makes one type of CNTs dominant [14]. While, these techniques are promising steps, they are still unreliable and the circuit design should take into account the residual presence of some metallic CNTs. Fault-tolerant design methodologies based on the statistical distribution of metallic CNTs were presented recently [5].

On the other hand, there have been some tries to improve the control of the misalignment of CNTs by using quartz as a substrate during the growth process [1, 15]. These techniques can be combined with robust design methodologies that account for misplaced CNTs and yield a reliable circuit design [16].

It was reported that CNTFETs outperform their CMOS counterpart in terms of speed and power in the ideal case, i.e., when metallic and misaligned CNTs are not present: the energy-delay product of CNTFETs is expected to be $\sim 13\times$ better than the value characterizing current MOSFETs at the 32 nm node [17]. Many approaches were conducted to transport existing MOSFET-based functions to CNT technology. This includes logic gates, memory and oscillators [18].

Different types of CNTFETs have been demonstrated in literature; the most important distinction is between MOSFET-like and *Shottky barrier CNTFETs* (*SB-CNTFETs*) [8]. While the first family is characterized by doped CNTs, the second family is made up of intrinsic CNTs that form a Shottky barrier at the drain and source contacts. SB-CNTFETs are ambipolar, i.e., they conduct both electrons and holes, showing a superposition of n- and p-type behaviors. The Shottky barrier thickness can be modulated by the fringing gate field at the CNT-to-metal contact; allowing the polarity of the device to be set electrically [6, 7]. A similar ambipolar behavior has been reported on graphenenanoribbon field-effect transistors, and suggests the possible electrical polarity control of these novel devices as well [19].

5.2 Ambipolar CNTFETs

While the uncontrollable ambipolar behavior is undesirable, the ability of controlling the CNTFET polarity (n- or p-type) in field by means of the fringing gate field raises the idea of using a second gate, the *polarity gate* (*PG*), to control the electrical field at the CNT/metal junction and to set the device polarity [6, 7]. The polarity gate is different from the usual gate, called the *control gate* (*CG*) in order to distinguish between both gates, since the PG applies an electrical field onto the

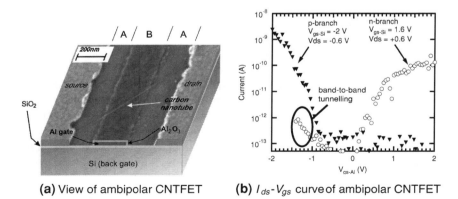

(a) View of ambipolar CNTFET   **(b)** I_{ds}-V_{gs} curve of ambipolar CNTFET

Fig. 5.1 Ambipolar CNTFET view and characterization [7]: **a** View based on a SEM of ambipolar double-gate CNTFET. Region A is back gate and B is top gate. **b** $I_{ds}-V_{gs}$ curve with top gate for a fixed back gate voltage. For a positive (negative) back gate voltage: device behaves as n- (p-) type

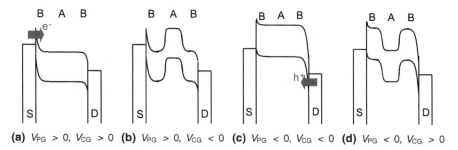

(a) $V_{PG} > 0$, $V_{CG} > 0$ **(b)** $V_{PG} > 0$, $V_{CG} < 0$ **(c)** $V_{PG} < 0$, $V_{CG} < 0$ **(d)** $V_{PG} < 0$, $V_{CG} > 0$

Fig. 5.2 Band diagram of an ambipolar CNTFET: **a** With $V_{PG} > 0$ the CNTFET behaves as a n-type device. For $V_{CG} > 0$, an electron current flows. **b** The n-type device blocks the electron current flow. **c** With $V_{PG} < 0$ the CNTFET behaves as a p-type device. For $V_{CG} < 0$, a hole current flows. **d** The p-type device blocks the hole current flow

CNT-to-drain and CNT-to-source junctions, while the CG applies an electrical field onto the CNT channel.

The ambipolar CNTFET used in [20] is based on the device demonstrated in [6, 7]. Figure 5.1 shows a view and the controllable characteristics of this ambipolar double-gate device. The top gate (control gate) in region A controls the current conduction through the device, while the back gate (polarity gate) in region B controls the type of polarity: a high or low voltage yield respectively an n- or p-type behavior. The working principle of these devices can be apprehended by means of the band diagram in Fig. 5.2. This device possesses a Schottky barrier at the drain and source contacts that can be thinned by applying the right contact in region B. If the voltage applied at the electrode controlling the region B is positive and large enough (V_+), then the Schottky barrier is transparent to tunneling electrons and the transistor has a n-type behavior. When the same voltage is negative and large enough (V_-), then the Schottky barrier is transparent to tunneling holes and the transistor has a p-type behavior. In [21, 22], it has been shown that between these two values, the barrier is too thick for both electrons and holes and the conduction through the transistor is poor, and it is minimal for a PG bias $V_0 = V_{ds}/2$ if V_{ds} is applied between drain and source. The conduction is also poor if the PG is left floating. While the choice of the voltage applied in region B determines the polarity of the devices, the voltage applied in region A may set up a high potential barrier in the middle of the channel and stop any potential current flow (Fig. 5.2).

From the technological point of view, the use of the substrate as a back gate operating as a PG means that all PGs are connected together. This reduces the opportunities for circuit design. It is possible to fabricate independent back-gates operating as PGs since the thermal budget for CNTFET technology is low. Then, the whole structure can be fabricated on top of CMOS during the process backend steps [23].

It may be also desirable to have both gates on top of the structures. This has not been demonstrated so far but it is feasible. Recently, a CNTFET with self-aligned

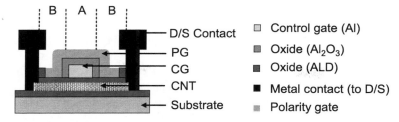

Fig. 5.3 Concept of self-aligned ambipolar CNTFET: metallic CG is defined first by lithography. PG is defined on top and isolated by the native oxide of CG. Drain and source metal contacts are opened and passivated at the PG edges

drain/source and gate has been presented in [24]. The goal of this work is to study the standard monopolar behavior of CNTFET. However, this self-alignment technique may be used in order to make PGs and CGs self-aligned. This yields a very compact device layout and enables its top gating. Figure 5.3 depicts a device cross-section with this potential approach. The main idea is the following: after the undoped CNTs are deposited on the oxidized substrate, they are covered by an *atomic layer deposition* (*ALD*) oxide. Then, an Aluminum control gate is patterned in the middle of the CNT. The native Al_2O_3 insures the insulation of the control gate from the polarity gate deposited on top of it. Then, vias through the terminals of the polarity gate enable the definition and the contacting of drain and source. The drain/source contacts are thereby isolated from the edges of the PG (for instance by a thin ALD oxide layer).

In anyone of the possible fabrication techniques of double-gate SB-CNTFETs, the device compact layout is depicted in Fig. 5.4a, showing 4 terminals corresponding to drain, source, PG and CG. The bulk terminal is not depicted because it is not used during normal operation. The bulk does not exist in the sense of MOSFETs since the CNT body is isolated from the Si substrate by means of a SiO_2 layer. However, a certain control of the device behavior can be performed by

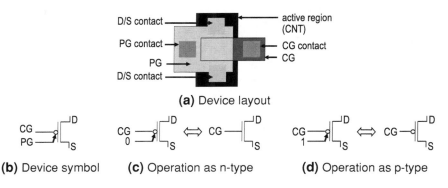

Fig. 5.4 Layout and symbol of ambipolar CNTFET: **a** Device layout. **b** Device symbol used in this chapter. **c** Device operation as n-type by setting PG to 0. **d** Device operation as p-type by setting PG to 1

using the Si substrate, which case is included in some CNTFET models [25]. The device symbol used in this chapter is depicted in Fig. 5.4b, and the operation as either n- or p-type device depending on the applied PG voltage is shown in Fig. 5.4c, d.

Ambipolar CNTFETs offer the opportunity of having *in-field programmable ambipolar devices*, i.e., devices whose n- or p-type behavior can be programmed in field by using the polarity gate. This novel feature of ambipolar CNTFETs has been investigated in [8], where a compact and in-field reconfigurable logic gate that maps 8 different logic functions of 2 inputs by using only 7 CNTFETs has been presented.

5.3 Dynamic Logic with Ambipolar CNTFETs

Dynamic logic has been shown to be an easy way to implement programmable logic functions with ambipolar CNTFETs [8]. In this section, dynamic logic is used to design an on-line reconfigurable NOR gate, called the *generalized NOR* (*GNOR*) gate.

A dynamic logic four-input NOR gate is presented in Fig. 5.5. While input B is inverted, inputs A and D are kept without inversion by setting the PG of the first, second and fourth CNTFET to V_+, V_- and V_+ respectively. In order to inhibit input C, the PG of the third CNTFET was set to V_0. As in usual dynamic logic, the transistors T_{PC} and T_{EV} are used to pre-charge and evaluate the output: These transistors have opposite polarities and they are clocked with non-overlapping complementary signals ϕ and $\overline{\phi}$: during the pre-charge phase, T_{PC} is conducting and T_{EV} is high resistive, so that Y is set high. During the evaluation phase, T_{EV} is conducting and T_{PC} is high resistive. Then, Y is set low if any one of the signals A, $\overline{B}$ or D is high. Consequently, this logic gate performs the function $\overline{A + \overline{B} + D}$.

The polarity of the inputs in the previously shown example is controlled by their respective polarity control signals. Let $V_+ \equiv$ logic-0 (n-type device) and $V_- \equiv$ logic-1 (p-type device). Then, every CNTFET implements a XOR operation between PG and CG logic signals. This is a very interesting feature of the

Fig. 5.5 Dynamic NOR gate with ambipolar CNTFETs:
$Y = \overline{A + \overline{B} + D}$

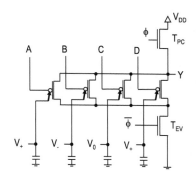

considered devices, since the XOR operation is implemented with one single transistor. Instead of fixing the PG signals, one can use them as additional inputs to the dynamic NOR gate, resulting in a dynamic GNOR gate, which efficiently combines NOR and XOR logic functions. The dynamic GNOR gate presented in Fig. 5.6 implements the function $Y = \overline{A \oplus B + C \oplus D}$ with a relatively small number of transistors, and makes use of the signals B and D as free variables. The transistors T_{PC} and T_{EV} execute the usual "precharge" and "evaluate" operations in dynamic logic.

The first interesting feature of the proposed design is its ability to include the XOR function in a compact way, which can be seen as an on-line programmable input inversion, because $A \oplus 1 = \overline{A}$ and $A \oplus 0 = A$. The second interesting feature of this design is the ability to select the inputs to be included in the computation: whenever the PG is left floating or set to V_0, the input of the concerned devices is not included in the computation, i.e., the device operates as an open circuit that does not influence the operation of the logic gate. Both advantages will be combined and used to design PLAs in Sect. 5.6.

This logic gate has two major weaknesses. On one hand, it is based on dynamic logic that is vulnerable to internal signal races. On the other hand, if both signals B and D are equal to 1, then the pull-down network will be formed exclusively by p-type devices. This can pull down the output to $\sim V_{SS} + |V_{Tp}|$ at most. The output does not reach the full swing and it worsens further when stages are cascaded, seriously compromising noise margins. The approach presented in the next section tries to resolve this problem by defining a static logic design for ambipolar CNTFETs.

5.4 Static Logic with Ambipolar CNTFETs

Dynamic logic design of ambipolar CNTFET gates is compact and easy to implement. However, it comes with all drawbacks of dynamic logic, namely signal racing and the difficult implementation of signal inversion, unless the circuit is made more complex [26]. Moreover, the output does not reach the output swing and necessitates the utilization of restoration stages. Static logic ambipolar

Fig. 5.6 Dynamic GNOR gate: $Y = \overline{A \oplus B + C \oplus D}$

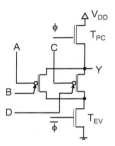

CNTFET gates can therefore represent an attractive design option. Signal racing and signal inversion do not represent any challenge for static logic. However, the availability of outputs with full swing is not ensured if ambipolar CNTFETs are configured as n-type devices in the *pull-up* (*PU*) network or as p-type devices in the *pull-down* (*PD*) network. The goal of this section is to implement circuit-level design styles in order to circumvent this situation.

5.4.1 Transmission-Gate Static Logic Family

The design style is improved by proposing two innovations with respect to the previous ambipolar CNTFET design approaches. The first innovation is that analogous to CMOS gates, full swing can be restored by inserting a PU network, that represents the complement of the PD network. This would cause a degradation of the output signal due to the potential presence of n-type (p-type) CNTFET(s) in the PU (PD) network. In fact, an n-type device in the PU network passes $V_{DD} - V_{Tn}$ at most, and a p-type device in the PD network passes $V_{SS} + |V_{Tp}|$ at least, causing in either cases a signal degradation. In order to avoid these configurations, we replace each CNTFET whose polarity is to be set during operation time by a transmission-gate formed by two CNTFETs controlled (at both the regular gate and the control gate) by complementary signals. In a transmission-gate, both n- and p-type devices are in parallel. If one of the devices fails in passing the full signal, the other device necessarily restores the signal level (Fig. 5.7).

The second innovation is to extend the GNOR gates to *generalized NAND* (*GNAND*) and *generalized AOI* (*GAOI*) configurations, by considering series and series-parallel combinations of transistors in the PU/PD paths. Figure 5.8 illustrates the circuit implementation of all gates that can be obtained this way by using no more than two transmission-gates and transistors in series in either PU or PD networks. The derivation of transistor aspect ratios (W/L), indicated in the figure, will be explained in Sect. 5.5.1.

With no more than three transmission-gates and transistors in series in either PU or PD networks, we obtain 46 different logic gates, which are listed in Table 5.1, with a maximum of three inputs (applied to the gates) and three control inputs (applied to the polarity gates). Even though every transmission-gate has two

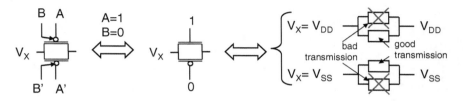

Fig. 5.7 CNTFET transmission-gate: any passing configuration $(A \oplus B = 1)$ prevents signal degradation

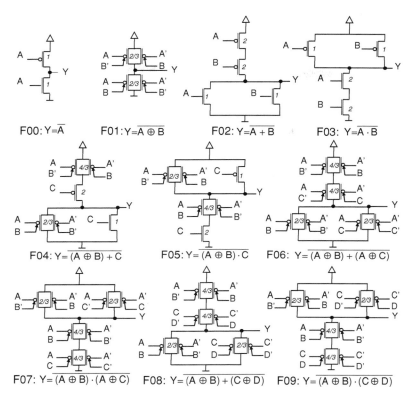

Fig. 5.8 Circuit implementation of ambipolar CNTFET logic gates with no more than 2 series transmission-gates or transistors in the PU/PD networks

transistors, a topologically uniform comparison between CNTFET- and CMOS-based gates suggests that we consider CMOS gates with three inputs at most, instead of six. Then, with the same constraints and topology, we obtain only 7 CMOS-based logic gates (F00, F02, F03, F10, F11, F16 and F17), highlighting the higher expressive power of the proposed transmission-gate-based static logic family.

In this design approach, whenever the function $U \oplus V$ is implemented with transmission-gate CNTFET, both polarities of U and V are needed, as illustrated in Fig. 5.8. By swapping the order in which the signals with different polarities are applied to the transmission-gates, it is possible to implement $\overline{U} \oplus V, U \oplus \overline{V}$ and $\overline{U} \oplus \overline{V}$. Since $U \oplus V = \overline{U} \oplus \overline{V}$ and $\overline{U} \oplus V = U \oplus \overline{V}$, it is possible to implement one more function by utilizing the same resources. For instance, the circuit implementing F05: $\overline{(A \oplus B) \cdot C}$ also implements $\overline{(\overline{A} \oplus B) \cdot C}$ by swapping the inputs A and $\overline{A}$. Note that in Sect. 5.5.4, the technology mapping tool is aware of the existence of additional gates obtained by swapping signal polarities.

Table 5.1 Ambipolar CNTFET logic gates with no more than 3 series transmission-gates or transistors in each PU/PD network

Gate	Function
F00	$\overline{A}$
F01	$\overline{A \oplus B}$
F02	$\overline{A + B}$
F03	$\overline{A \cdot B}$
F04	$\overline{(A \oplus B) + C}$
F05	$\overline{(A \oplus B) \cdot C}$
F06	$\overline{(A \oplus B) + (A \oplus C)}$
F07	$\overline{(A \oplus B) \cdot (A \oplus C)}$
F08	$\overline{(A \oplus B) + (C \oplus D)}$
F09	$\overline{(A \oplus B) \cdot (C \oplus D)}$
F10	$\overline{A + B + C}$
F11	$\overline{(A + B) \cdot C}$
F12	$\overline{(A \oplus D) + (B \oplus D) \cdot (C \oplus D)}$
F13	$\overline{((A \oplus D) + B) \cdot C}$
F14	$\overline{((A \oplus D) + (B \oplus D)) \cdot C}$
F15	$\overline{((A \oplus D) + (B \oplus D)) \cdot (C \oplus D)}$
F16	$\overline{A + B \cdot C}$
F17	$\overline{A \cdot B \cdot C}$
F18	$\overline{(A \oplus D) + B + C}$
F19	$\overline{(A \oplus D) + (B \oplus D) + C}$
F20	$\overline{(A \oplus D) + (B \oplus D) + (C \oplus D)}$
F21	$\overline{(A \oplus D) + B \cdot C}$
F22	$\overline{A + (B \oplus D) \cdot C}$
F23	$\overline{(A \oplus D) + (B \oplus D) \cdot C}$
F24	$\overline{A + (B \oplus D) \cdot (C \oplus D)}$
F25	$\overline{(A \oplus D) \cdot B \cdot C}$
F26	$\overline{(A \oplus D) \cdot (B \oplus D) \cdot C}$
F27	$\overline{((A \oplus D) + B) \cdot (C \oplus D)}$
F28	$\overline{(A + B) \cdot (C \oplus D)}$
F29	$\overline{A + (B \oplus D) \cdot (C \oplus E)}$
F30	$\overline{((A \oplus D) + B) \cdot (C \oplus E)}$
F31	$\overline{(A \oplus D) \cdot (B \oplus D) \cdot (C \oplus D)}$
F32	$\overline{(A \oplus D) + (B \oplus E) + C}$
F33	$\overline{(A \oplus D) + (B \oplus D) + (C \oplus E)}$
F34	$\overline{(A \oplus D) + (B \oplus E) \cdot C}$
F35	$\overline{(A \oplus D) + (B \oplus E) \cdot (C \oplus D)}$
F36	$\overline{(A \oplus D) + (B \oplus E) \cdot (C \oplus E)}$
F37	$\overline{(A \oplus D) \cdot (B \oplus E) \cdot C}$
F38	$\overline{(A \oplus D) \cdot (B \oplus D) \cdot (C \oplus E)}$

(continued)

Table 5.1 (continued)

Gate	Function
F39	$\overline{(A \oplus D) + (B \oplus E) + (C \oplus F)}$
F40	$\overline{(A \oplus D) + (B \oplus E) \cdot (C \oplus F)}$
F41	$\overline{(A \oplus D) \cdot (B \oplus E) \cdot (C \oplus F)}$
F42	$\overline{((A \oplus D) + (B \oplus D)) \cdot (C \oplus E)}$
F43	$\overline{((A \oplus D) + (B \oplus E)) \cdot C}$
F44	$\overline{((A \oplus D) + (B \oplus E)) \cdot (C \oplus D)}$
F45	$\overline{((A \oplus D) + (B \oplus E)) \cdot (C \oplus F)}$

5.4.2 Alternate CNTFET Families

Some alternate CNTFET families with less transistor counts can be derived from the transmission-gate static logic family defined before. In the first approach, the transistor count can be reduced by replacing the PU network by a single PU transistor, resulting in a pseudo logic style. The PU CNTFET is weaker than the PD devices in order to allow the output signal to fall within the tolerated noise margin. The gates are expected to be slower because of the weak pull-up. Higher static power is also a potential concern. The pseudo logic implementation of the same set of logic functions listed in Table 5.1 can be derived, as illustrated in Fig. 5.9a for F05.

The second approach to reduce transistor count is to replace all transmission-gates by pass-transistors, in static or pseudo logic configurations. Figure 5.9b, c illustrate the pass-transistor implementations of F05, as an example, in static and pseudo logic styles respectively. However, this implies that CNTFETs that are electrically configured as n- or p-type can be located in the PU or PD network, respectively. Since this may degrade the output level, a restoration stage (inverter) is used to restore full swing at the output. The area-delay costs of this approach are assessed in Sect. 5.5.3.

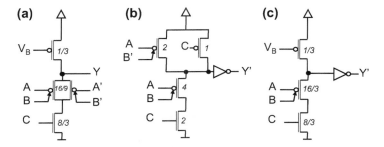

Fig. 5.9 Compact implementation of F05: $\overline{(A \oplus B) \cdot C}$: transmission-gate pseudo logic **a**, pass-transistor static logic **b** and pass-transistor pseudo logic **c**

Actually, all these alternate families represent static logic. But the static complementary logic (including both transmission-gates and pass-transistors) is simply called static logic family, and the static pseudo logic (including both transmission-gates and pass-transistors) is simply called pseudo logic family. This is an implicit way of naming these families that is frequently found in literature.

5.5 Multi-Level Logic Synthesis with Static CNTFET Gates

We designed the static logic gates such that the rise and fall times are identical, and the output current is equal that of the unit inverter. We took into account the equal on-resistance of n- and p-type CNTFETs with the same size, due to equal electron and hole mobility in CNTs. Thus, unlike MOSFET gates, the pull-up devices in CNTFET gates need not be larger than the pull-down devices in order to achieve an equal on-resistance. This yields smaller CNTFET gates compared to the MOSFET gates implementing the same function.

We simulated the correct operation of the designed CNTFET families with the Stanford CNTFET model for unipolar devices [25], using a lithography pitch of 32 nm. At the time this book is written, no SPICE model for controllable ambipolar CNTFETs has been released. We therefore fixed the polarity gate signals, i.e., the device polarities during simulations in a similar way to [8]; and we used parallel n- and p-type devices in order to simulate the ambipolar behavior.

In the following, details of the design of the transmission-gate static logic family and other alternate static logic families are given. Then, the designed libraries are characterized in terms of speed and area. The characterized libraries are utilized for multi-level synthesis of a benchmark of logic circuits.

5.5.1 Transmission-Gate Static Design

We denote by R_n (R_p) the on-resistance of the n-type (p-type) device. The resistance of a transistor passing a signal in the wrong direction is roughly double its on-resistance [27]. Hence, the resistance of a transmission-gate is estimated as $R_n \parallel 2R_p$ if it conducts a low signal, and $2R_n \parallel R_p$ if it conducts a high signal. Since $R = R_n = R_p$ holds for CNTFETs, the equivalent resistance of the transmission-gate is always equal to $\sim 2R/3$. These values were taken into account in sizing the transmission-gates. The use of transmission-gates have opposite effects. On the one hand, the decrease of the on-resistance to $\sim 2R/3$ instead of R speeds up the gate. On the other hand, transmission-gates with a unit on-resistance have a larger area ($2 \times 2A/3$) than unit transistors (A). Thus, they have a higher input capacitance that may slow the gate, and they are more expensive in terms of area.

5.5.2 Design of Alternative CNTFET Families

Similarly, the pass-transistors were sized in order to achieve an equal rise and fall time and to drive as much current as a unit inverter. Since the pass-transistors are potentially passing in the wrong direction (n-type in PU or p-type in PD network), their worst-case on-resistance is $2R$. Thus, they were designed to be double the unit size (area = $2A$). Despite the reduction in transistor count of the pass-transistor family compared to the transmission-gate family, the cost in terms of area to achieve a unit on-resistance is larger ($2A$ vs. $4A/3$). Consequently, transmission-gates are preferable to pass-transistors in static logic. In pseudo logic, pass-transistors may be useful because the logic gates require no inverted inputs, unlike other logic families. We assumed for pseudo logic gates (with either transmission-gates or pass-transistors) that the PU device is $4\times$ weaker than the PD network, which is a good compromise between delay and area.

5.5.3 Library Characterization

Table 5.2 summarizes the area and FO4 delay estimates for the library cells. Note that the additional gates obtained by swapping the signal polarities at the transmission-gates (Sect. 5.4.1) have the same area and delay as the gates from which they were derived. Then, we compared them to their MOSFET counterpart, whenever they exist with *the same topology*; i.e., with no more than 3 series transistors in the PU and PD networks respectively. The area of the logic gates was estimated in a normalized way as the number of transistors multiplied by their respective size ratio (W/L); all the gates being designed such that they drive the current of a unit inverter. The *fan-out-of-4 (FO4)* delay was calculated with the switch-level RC delay model [27] and is equal to the delay of a gate driving 4 instances of itself. In this model, the FO4-delay is given by $p + 4g$, where p is the parasitic (or intrinsic) delay of the logic gate (delay to charge the gate parasitic output capacitances with no load), and g is the logical effort (the ratio of the gate input capacitance to the input capacitance of the inverter delivering the same current) [27]. The input capacitance of the polarity gate and the actual gate were assumed to be equal. We also assumed for CNTFETs the common approximation for MOSFETs, which consist in considering that the gate capacitance is almost equal to the drain/source parasitic capacitances. We calculated the FO4 delay in average (for all inputs) and in the worst case (for the slowest input). The FO4 delay was normalized to the delay of a unit inverter τ (defined as the delay of a fanout-of-1 inverter with no parasitic capacitances), which is technology-dependent, and for CNT technology it is about $5.1\times$ better than CMOS [17].

We noticed that the static transmission-gate XNOR gate has a better FO4 than the unit inverter. This is due to the lower parasitic drain capacitance of the transmission-gates (in XNOR) compared to an inverter driving the same output current. Most of the cells designed with static transmisson-gates present this advantage. Thus, the normalized average FO4 delay of all CNTFET transmission-gate static logic gates is equal to the one of all static CMOS gates, even though the implemented functions are by far more complex. On the other hand, the unique property of CNTFET that results in the same on-resistance of equally sized p- and n-type devices makes the CNTFET cells more compact: despite the larger average number of transistors in the CNTFET static library compared to static CMOS, its average area is slightly smaller (12.3 vs. 12.7). As expected, the CNTFET transmission-gate pseudo logic family has a 31% smaller average gate area than its static counterpart (8.5 vs. 12.3); however, it is 33% slower (12 vs. 9). Surprisingly, the CNTFET pass-transistor pseudo logic family is less area efficient than its transmission-gate counter-part. This confirms the conjecture in Sect. 5.5.2 that larger area is needed for pass-transistors in order to compensate for the high on-resistance of p-type (n-type) transistors operating in the PD (PU) network. This family is only 7% more compact than the transmission-gate static logic family (average area: 11.5 vs. 12.3), while it is 2.7× slower (9 vs. 24.1). This makes the CNTFET pass-transistor family a bad choice in circuit design. The CNT logic families explored here need both polarities of inputs involved in XOR operations. Consequently, we included an output inverter in every gate, in order to provide both polarities of every output (same Table 5.2).

5.5.4 Logic Synthesis Results

The tool ABC developed at Berkeley [28] was used at Rice University to perform synthesis and technology mapping of several benchmark circuits.[1] The circuits were first synthesized using the resyn2rs script, followed by technology mapping using libraries for each family based on the area-delay values from Table 5.2. The results for 15 benchmark circuits are summarized in Table 5.3. In Sects. 5.4.2 and 5.5.3, we demonstrated that the transmission-gate configuration outperforms the pass-transistor configuration in terms of area and delay. We therefore considered only transmission-gate implementations in static and pseudo logic and we compared them with a CMOS mapping. For each family, the number of gates, the normalized circuit area (to a unit transistor), the logic depth, the normalized delay (to the technology-dependent intrinsic delay τ [17]), and the absolute delay in picoseconds are reported. Whereas both CNTFET families reduce the implementation complexity, the static family is more efficient in terms of speed and the pseudo family is more attractive in terms of area.

The trend of some functions in terms of area and delay and the average tendency are depicted in Fig. 5.10: circuits that embed XOR operations extensively—the

---

[1] The multi-level logic synthesis reported here was fully performed by Prof. Mohanram's group at Rice University, TX, USA.

Table 5.2 Characterization of the designed CNTFET library compared to CMOS *with the same topology*: transistor count (T), normalized area (A) to a unit transistor, normalized FO4 delay to the technology-dependant delay τ [17] in the worst case (w) and on average (a). Average performance of gates without and with output inverters are also indicated

| Gate | CNTFET technology | | | | | | | | | | | | CMOS technology | | | |
| | Transmission-gate static logic | | | | Transmission-gate pseudo logic | | | | Pass-transistor pseudo logic | | | | Static logic | | | |
	T	A	$\frac{FO4}{\tau_1}$(w)	$\frac{FO4}{\tau_1}$(a)	T	A	$\frac{FO4}{\tau_1}$(w)	$\frac{FO4}{\tau_1}$(a)	T	A	$\frac{FO4}{\tau_1}$(w)	$\frac{FO4}{\tau_1}$(a)	T	A	$\frac{FO4}{\tau_2}$(w)	$\frac{FO4}{\tau_2}$(a)
F00	2	2	5	5	2	1.7	7	7	2	1.7	7	7	2	2	5	5
F01	4	2.7	4	4	3	2.1	5.7	5.7	2	3	13.7	13.7	–	–	–	–
F02	4	6	8	8	3	3	8.3	8.3	3	3	8.3	8.3	4	10	8.7	8.7
F03	4	6	8	8	3	5.7	13.7	13.7	3	5.7	13.7	13.7	4	8	7.3	7.3
F04	6	7	8.2	6.6	5	3.4	8.8	7.4	3	4.3	15	13.2	–	–	–	–
F05	6	7	8.2	6.6	5	6.6	13.7	10.8	3	13.7	27	23.4	–	–	–	–
F06	8	8	10.7	8	5	3.9	11	8.6	3	5.7	27	19.9	–	–	–	–
F07	8	8	10.7	8	5	7.4	18.1	13.4	3	11	48.3	34.1	–	–	–	–
F08	8	8	6.7	6.7	5	3.9	7.4	7.4	3	5.7	16.3	16.3	–	–	–	–
F09	8	8	6.7	6.7	5	7.4	11	11	3	11	27	27	–	–	–	–
F10	6	12	11	11	4	4.3	9.7	9.7	4	4.3	9.7	9.7	6	21	12.3	12.3
F11	6	11	10.5	9.8	4	8.3	13.7	13.7	4	8.3	13.7	13.7	6	16	10.7	9.8
F12	12	14.7	18	10.7	7	9.2	23.4	14.6	4	7	31	17.7	–	–	–	–
F13	8	12.3	10.5	8.4	5	9.2	13.7	11.3	4	11	24.3	20.8	–	–	–	–
F14	10	13.7	13.5	9.8	6	10.1	17.2	12.7	4	13.7	45.7	28.9	–	–	–	–
F15	12	14.7	18	10.7	7	11	25.2	14.6	4	16.3	69.7	37.7	–	–	–	–
F16	6	11	10.5	9.8	4	7	15	13.2	4	7	15	13.2	6	17	10.3	9.9
F17	6	12	11	11	4	12.3	20.3	20.3	4	12.3	20.3	20.3	6	15	9.7	9.7
F18	8	13.3	11.2	9.4	5	4.8	10.1	8.9	4	5.7	16.3	13.7	–	–	–	–
F19	10	14.7	11.3	10.6	6	5.2	12.3	10.1	4	7	28.3	19	–	–	–	–
F20	12	16	20	12	7	5.7	16.3	11	4	8.3	40.3	24.3	–	–	–	–

(continued)

Table 5.2 (continued)

Gate	CNTFET technology								Pass-transistor pseudo logic				CMOS technology			
	Transmission-gate static logic				Transmission-gate pseudo logic								Static logic			
	T	A	$\frac{FO4}{\tau_1}$(w)	$\frac{FO4}{\tau_1}$(a)	T	A	$\frac{FO4}{\tau_1}$(w)	$\frac{FO4}{\tau_1}$(a)	T	A	$\frac{FO4}{\tau_1}$(w)	$\frac{FO4}{\tau_1}$(a)	T	A	$\frac{FO4}{\tau_2}$(w)	$\frac{FO4}{\tau_2}$(a)
F21	8	12	11	8.3	5	7.4	15.4	10.7	4	8.3	16.3	16.3	–	–	–	–
F22	8	12.3	10.5	8.4	5	7.9	13.7	10.4	4	9.7	25.7	19	–	–	–	–
F23	10	13.3	12.3	9.5	6	7	15.4	12.4	4	11	37.7	24.3	–	–	–	–
F24	10	13.7	13.5	9.8	6	8.8	26.6	14.1	4	12.3	49.7	29.7	–	–	–	–
F25	8	13.3	11.2	9.4	5	13.7	20.3	16.8	4	16.3	36.3	28.3	–	–	–	–
F26	10	14.7	14	10.6	6	15	20.3	10.7	4	20.3	68.3	40.3	–	–	–	–
F27	10	13.3	12.3	10.1	6	10.1	18.1	13.5	4	13.7	48.3	31.6	–	–	–	–
F28	8	12	11	8.3	5	9.2	14.6	12.2	4	11	27	23.4	–	–	–	–
F29	10	13.7	10.8	8.5	6	10.1	13.7	10.5	4	13.7	24.3	13.2	–	–	–	–
F30	10	13.3	11	8	6	10.1	14.6	11.4	4	13.7	27	25.8	–	–	–	–
F31	12	16	20	12	7	16.3	37.7	21.7	4	24.3	104.3	56.3	–	–	–	–
F32	10	14.7	11.3	11	6	5.2	14.1	12.5	4	7	17.7	16.6	–	–	–	–
F33	12	16	14.7	10.4	7	5.7	12.8	9.3	4	8.3	29.7	21.1	–	–	–	–
F34	10	13.3	11	8	6	8.3	15.4	10.7	4	11	27	20.6	–	–	–	–
F35	12	14.7	12.7	9.2	7	9.2	16.3	12.8	4	13.7	40.3	29.7	–	–	–	–
F36	12	14.7	14	9.2	7	9.2	19.9	12.8	4	13.7	51	29.7	–	–	–	–
F37	10	14.7	11.3	9	6	15	20.3	15.6	4	20.3	36.3	33.1	–	–	–	–
F38	12	16	14.7	10.4	7	16.3	27	18.5	4	24.3	72.3	46.7	–	–	–	–
F39	12	16	9.3	9.3	7	5.7	9.2	9.2	4	8.3	19	19	–	–	–	–
F40	12	14.7	8.7	8.2	7	9.2	12.8	11.6	4	13.7	29.7	26.1	–	–	–	–
F41	12	16	9.3	9.3	7	16.3	16.3	16.3	4	24.3	40.3	40.3	–	–	–	–
F42	14	12.7	14	9.2	7	11	18.1	12.4	4	16.3	48	31.3	–	–	–	–
F43	10	13.7	8.8	8.2	6	10.1	13.7	10.5	4	13.7	24.3	23.2	–	–	–	–

(continued)

Table 5.2 (continued)

| Gate | CNTFET technology | | | | | | | | | | | | CMOS technology | | | |
| | Transmission-gate static logic | | | | Transmission-gate pseudo logic | | | | Pass-transistor pseudo logic | | | | Static logic | | | |
	T	A	$\frac{FO4(w)}{\tau_1}$	$\frac{FO4(a)}{\tau_1}$	T	A	$\frac{FO4(w)}{\tau_1}$	$\frac{FO4(a)}{\tau_1}$	T	A	$\frac{FO4(w)}{\tau_1}$	$\frac{FO4(a)}{\tau_1}$	T	A	$\frac{FO4(w)}{\tau_2}$	$\frac{FO4(a)}{\tau_2}$
F44	12	14.7	14	9.2	7	11	18.1	12.4	4	16.3	48.3	31.3	–	–	–	–
F45	12	14.7	8.7	9.2	7	11	11	11	4	16.3	32.5	24.1	–	–	–	9
Av. w/o INV	9.1	12.3	11.3	9	5.6	8.5	15.6	12	3.7	11.5	32.5	24.1	4.9	12.7	9.1	9
Av. w/ INV	11.1	14.3	–	10.7	7.6	10.2	–	13.8	5.7	13.1	–	25.5	–	–	–	–
τ	$\tau_1 = 0.59$ ps				$\tau_1 = 0.59$ ps				$\tau_1 = 0.59$ ps				$\tau_2 = 3.00$ ps			

Table 5.3 Technology mapping results: Gate count, normalized circuit area (to area of unit transistor), logic depth, normalized circuit delay (to technology-dependent intrinsic delay τ [17]) and absolute delay (in ps) for different benchmarks and technologies

Benchmark			CNTFET transmission-gate static logic					CNTFET transmission-gate pseudo logic					CMOS static logic				
			Gates	Area	Levels	Delay		Gates	Area	Levels	Delay		Gates	Area	Levels	Delay	
Name	I/O	Function	No.	Area	Levels	Norm.	Abs.	No.	Area	Levels	Norm.	Abs.	No.	Area	Levels	Norm.	Abs.
C2670	233/140	ALU and control	416	3292.5	12	105.2	62.1	467	1883.9	11	125.3	73.9	674	5687.0	16	120.0	360.0
C1908	33/25	Error correcting	201	1562.2	12	106.5	62.8	207	893.6	13	120.2	70.9	502	4641.0	22	175.0	525.0
C3540	50/22	ALU and control	642	6228.7	19	180.7	106.7	664	3475.4	19	197.6	116.6	956	8823.0	29	218.2	654
dalu	75/16	Dedicated ALU	679	6662.3	16	163.6	96.5	713	3956.8	17	193.5	114.2	1100	9181.0	28	205.9	617.7
C7552	207/108	ALU and control	904	6747.6	17	149.1	88.0	987	4235.7	17	174.4	102.9	1860	13933.0	24	173.6	520.8
C6288	32/32	Multiplier	1389	11672.9	48	397.8	234.7	1322	6558.0	48	481.6	284.1	2767	23192.0	89	639.8	1919.4
C5315	178/123	ALU and selector	894	7600.6	16	145.6	85.9	986	4553.2	17	172.2	101.6	1465	12048.0	27	200.2	600.6
des	256/245	Data encryption	2583	25781.1	10	88.1	52.0	2500	13920.0	9	90.8	53.6	3560	35781.0	15	115.3	345.9
i10	257/224	Logic	1279	11264.2	19	200.0	118.0	1287	6296.2	21	222.3	131.2	1965	16394.0	29	218.8	656.4
t481	16/1	Logic	670	6379.0	12	113.7	67.1	598	3516.0	11	114.0	67.3	804	8259.0	13	102.2	306.6
i18	133/81	Logic	674	6642.0	8	83.6	49.3	714	3698.6	9	89.8	53.0	836	7968.0	11	82.1	246.3
C1355	41/32	Error correcting	207	1260.2	9	63.9	37.7	215	776.6	9	73.6	43.4	579	5376.0	16	125.0	375.0
add-16	33/17	16-bit adder	128	834.4	19	179.2	105.7	132	540.0	20	220.0	129.8	217	1548.0	33	244.6	733.8
add-32	65/33	32-bit adder	256	1656.7	35	340.5	200.9	260	1091.4	36	421.6	248.7	441	3084.0	65	479.1	1437.3
add-64	129/65	64-bit adder	512	3321.0	67	663.1	391.2	516	2194.1	68	824.8	486.6	889	6156.0	129	948.3	2844.9
Average			762.3	6727.0	21.3	198.7	117.2	771.2	3839.3	21.7	234.8	138.5	1241.0	10804.7	36.4	269.9	809.7
Improvement versus CMOS			38.6%	37.7%	41.5%	26.4%	6.9×	37.9%	64.5%	40.4%	13.0%	5.8×	-	-	-	-	-
Delay normalization factor [17]			$\tau_1 = 0.59$ ps					$\tau_1 = 0.59$ ps					$\tau_2 = 3.00$ ps				

Fig. 5.10 Comparison of the mapping of chosen functions in the benchmark and the benchmark average between ambipolar CNT and CMOS technology. The improvement of CNT versus CMOS technology (area saving and reduction) is shown in percentage of the CMOS implementation. Comparisons are between normalized values

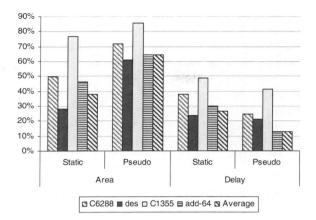

adders, ALUs, error correcting circuits, and the multiplier C6288—return the largest area and speed improvements when implemented in CNTFET technology.

The implementation with both transmission-gate CNT families requires on average $\sim 38\%$ fewer gates and 40% less logic levels than CMOS. While the static logic CNTFET family saves 37.7% area on average compared to CMOS, the pseudo logic CNTFET family saves 64.5% area on average. The area normalization factor was set to the area of a unit transistor, which is expected for MOSFET and ambipolar CNTFET to be equal [8], since the additional polarity gate is buried underneath the channel or defined on top of the actual gate. However, we may expect a negligible area cost due to the contact area of the polarity gate.

The circuits implemented in static and pseudo CNTFET families are 26.4% and 13.0% faster than the CMOS implantation respectively in terms of normalized delay. Delay was normalized to the technology-dependent intrinsic delay τ, which is expected to be $5.1\times$ faster for unipolar CNT technology than for CMOS [17]. We assumed the same intrinsic delay for unipolar and ambipolar CNTFETs and we calculated the absolute delay of the implemented circuits compared to the CMOS

Fig. 5.11 Ratio of the absolute delay of CMOS to CNTFET implementation

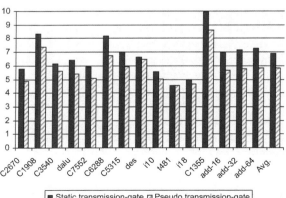

implementation. Figure 5.11 shows the cumulative benefits of technology and design that translate into an average speed-up of 6.9× and 5.8× for static and pseudo CNT families respectively compared to CMOS. The largest speed-up was calculated for the static CNT implementation of multipliers ($\sim 10\times$) and error correcting circuits (more than 8×). For delay calculations we considered the worst case scenario, in which every signal, i.e., either input or control signal, needs the charge or discharge an input capacitance equal to a unit drain/source intrinsic capacitance with every switching operation. Consequently, the presented estimation of the delay is the worst-case value. Even though the delay due to signal routing around ambipolar cells was not considered, it is expected to be canceled out by the benefit of smaller cell layout.

The designed logic gates ultimately need no additional internal signals, they just implement the same functions in a more compact way compared to CMOS.

5.6 Design of Regular Fabrics

A regular fabric is a set of resources (logic gates, memory, interconnect...) laid out in a regular manner, and that can be mask- or in-field configured to implement specific logic functions. Various forms of regular gate and logic arrays have been recently proposed to reduce the design risk caused by the increasing variability at the current and future technology nodes [29–33].

The designed logic gates in both dynamic and static logic can be efficiently organized into regular circuits that can be on-line re-configured. This organization is reminiscent to an FPGA architecture, but it uses logic functions instead of LUTs inside the *complex logic blocks (CLBs)*. This section introduces two novel design approaches for regular fabrics with ambipolar CNTFETs in both dynamic and static logic.

5.6.1 Dynamic PLA Architecture

Dynamic GNOR gates can be integrated into an array-based architecture, which is reminiscent of the regular PLA (Fig. 5.12). This architecture consists of a cascade of two planes, each implementing the GNOR function (Fig. 5.13). The depicted GNOR PLA plane is just an extension of the GNOR gate presented in Sect. 5.3 to an arbitrary number of inputs. In order to avoid the use of an additional wire per CNTFET for every PG signal, a charge corresponding to the voltage of the wished polarity is saved on every PG. A global signal V_{PG} connects all the polarity gates. Any transistor in position (i, j) whose polarity is to be set is selected by using the row and column select signal $V_{SelR,i}$ and $V_{SelC,j}$. During the configuration phase of the PLA, every ambipolar CNTFET is selected individually and the charge corresponding to its PG voltage is set. This insures an individual programming of every device.

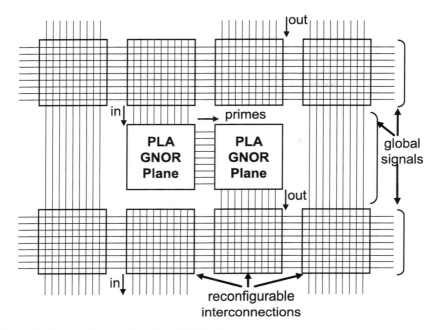

Fig. 5.12 PLA architecture based on GNOR planes

A compact interconnect array can be realized by using ambipolar CNTFET: every crosspoint connects a horizontal and a vertical wire through a CNTFET working as a pass transistor (Fig. 5.14). All CG voltages are set at the same high level. If the PG of the CNTFET is set to V_+, then the polarity of the CNTFET is n. The high level of CG makes the device conducting; then the wires are connected. If the PG of the CNTFET is set to V_-, then the device polarity is set to p. It is therefore switched off because CG is at a high level, and the wires are disconnected. Interleaving PLA and interconnects (Fig. 5.12) enables cascades of NOR planes and realizes any logic function.

Classical PLA planes need both polarities of input signals, whereas the use of GNOR gates prevents the replication of input columns. This powerful feature potentially reduces the size of the PLA even if the size of the basic cell is large. This area was estimated from the scaling rules suggested in [16] for CNTFET. The area of Flash and EEPROM basic cells were derived from the International Technology Roadmap for Semiconductors. The area of the contacted cells with respect to the lithography resolution (f) is estimated in the first row of Table 5.4.

The area of the PLA implementing three functions from the MCNC suite [34] is shown in Table 5.4. The CNTFET basic cell is approximately 50% larger than the Flash and 40% smaller than the EEPROM basic cell. PLAs based on ambipolar CNTFETs need only one polarity of every input, while PLAs with the same organization, but implemented in Flash or EEPROM technology, necessitate the

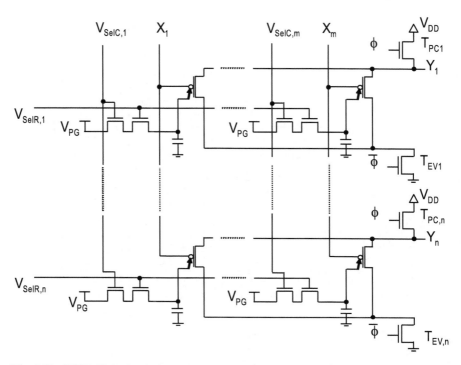

Fig. 5.13 GNOR PLA plane with ambipolar CNTFET

existence of both polarities of inputs, because signals cannot be inverted locally in these technologies, unlike in ambipolar CNTFET technology.

Then, the CNTFET PLA is always more compact than EEPROM PLA (up to 68% less area). Because Flash basic cells are smaller than ambipolar CNTFET cells, the CNTFET implementation can only save area compared to Flash if the PLA has a large number of inputs (e.g. in max46: saving $\sim 21\%$), by taking advantage of its fewer inputs; otherwise a small area overhead (3%) can be seen.

For PLA-based FPGA, this reduction in area is highly desirable because it facilitates the routing of signals between the CLBs. Moreover the number of signals to route is reduced by almost the factor 2, because the inverted signals are not routed but generated internally. These factors boost the performance of the routing tool. Consequently, the delay, which highly depends on signal routing in FPGA, can be drastically reduced.

Another advantage in using GNOR gates is the availability of the product-terms (output of first plane) with both polarities, thus allowing for a further degree of freedom in minimizing the PLA. A logic minimizer was presented in [35] and implemented in the heuristic MINI II, showing a significant area saving after logic minimization. The cascade of 4 NOR plane instead of 2 makes the implementation of *whirlpool PLAs* (*WPLAs*) [36] with the presented architecture possible. WPLAs outperform other PLA types and a more compact implementation can be obtained

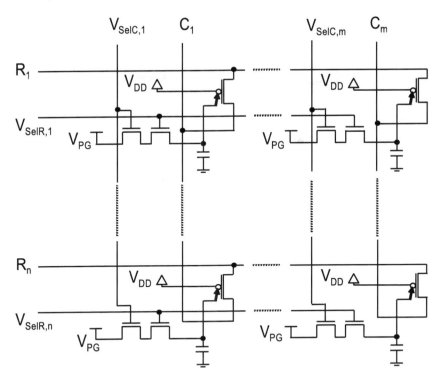

Fig. 5.14 Reconfigurable interconnect plane with ambipolar CNTFET

by using the logic minimizer called Doppio-Espresso [36]. Finally, a fault-tolerant design approach for PLAs [37] makes use of the regular architecture and is expected to improve the yield of the unreliable devices making up the PLA.

5.6.2 Static Regular Fabrics

The presented CNTFET logic gates in static logic have a higher expressive power than their MOSFET counterparts (Sect. 5.4). Their regular structure motivates

Table 5.4 Area of 3 logic functions implemented in different technologies and normalized to f^2, with f the lithography half-pitch

	Flash	EEPROM	CNTFET	CNTFET versus Flash (%)	CNTFET versus EEPROM (%)
Basic cell	40	100	60	+50	−40
max46	34960	87400	27600	−21	−68
apla	32000	80000	33000	+3	−59
t2	104000	260000	102960	−1	−60

Fig. 5.15 Baseline architecture of an ambipolar CNTFET regular fabric in static logic **a** and type 1 and type 2 logic blocks **b** and **c**

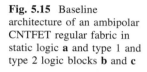

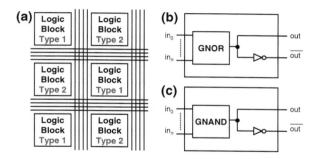

their use to design regular fabrics. The baseline architecture of a static ambipolar CNTFET regular fabric is depicted in Fig. 5.15a. Two types of logic blocks are interleaved. Their respective outputs are routed throughout the circuit by means of an interconnection network, which can be configured with SRAM cells in a similar way to MOSFET Field Programmable Logic Arrays (FPGA).

A more detailed view into the two types of logic blocks is illustrated in Fig. 5.15b, c. The main components of the logic blocks are generalized NOR and NAND gates whose circuit implementation with CNTFET technology is presented in Fig. 5.16. The design takes advantage of their identical physical layout rotated by 180°. Depending on the signals connected to the inputs of the generalized gates, they can be configured in order to implement a large set of cells from the library presented in Sect. 5.4.1. The design of the generalized gates in the other logic families can be derived in a straightforward way from the static transmission-gate family.

The advantage of an in-field programmable regular fabric is the simplicity of the design flow, the reconfigurability of the circuit and the immunity to process variability. It also offers the opportunity to estimate an upper bound on the delay of each stage, because the regular and symmetrical design of the generalized gates causes the FO4 delay to be the same in almost all the cases. If the block delay is known *a priori*, and the local routing delay is small enough, then the gates can be

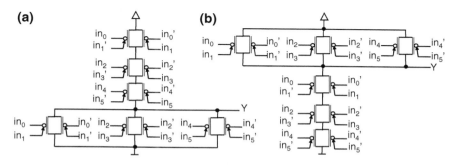

Fig. 5.16 Circuit implementation of GNOR **a** and GNAND **b** gates for ambipolar CNTFET regular fabrics in static logic

designed with dynamic logic with no risk of internal signal races, because the clock signals for the evaluation and pre-charge transistors can be delayed according to the maximum expected delay of the logic blocks. This yields a more robust dynamic logic, while taking advantage of its lower power and area compared to static logic.

5.7 Dicussions

This part of the work deals with a new variant of CNTFET technology, and it represents a first assessment of the benefits of this immature technology in terms of enhancements of the design abilities. The technology is still young: in addition to all challenges characterizing standard CNTFET technology, i.e., doped unipolar CNTFETs, the considered technology has its own challenges, such as the technology integration of the second gate and the ability to electrostatically control the CNT-to-metal junction with the additional gate. The physics of this 1D to 3D contact is actually not fully understood, which makes the device modeling more difficult. Ambipolar CNTFET technology needs to solve all these problems, in addition to the inherent problems of any CNTFET technology, such as bad controllability of the CNT diameter, chirality and alignment. Once all these problems are solved, the demonstrated benefits of the technology can be considered in logic design.

It is also important to be aware of another source of uncertainty: even if the technology becomes mature and stabilized, the devices will have some properties that fundamentally differ from MOSFET devices. As a matter of fact, the current through ambipolar CNTFET is a dominated by a tunnel current, unlike the drift-diffusion current in MOSFETs. Consequently, it has been noticed that ambipolar devices with undoped channel have a V_{ds}-dependent I_{off}. This remark concerns also FETs with graphene nanoribbon or intrinsic SiNW channels. The drawback of this property is that I_{off} increases exponentially when V_{ds} decreases, which may be a source of a high leakage. On the other hand, the tunnel current is not expected to saturate; i.e., I_{ds} does not saturate for large V_{ds} and a fixed V_{gs}. This means that additional circuit design techniques may be needed to address these issues.

When it comes to the electrical properties of ambipolar CNTFET devices, it is important to notice that the voltage range for CGs and PGs are not necessarily the same, as implicitly assumed in this chapter. Actually, in the demonstrated devices [7], the range is different. The problem can be solved at two different levels. If there is a margin for some extra technological choices, then, the respective work function of both gates as well as their respective oxide thickness can be engineered in order to match the voltage range of both types of gates. If this solution is not available, then the designer has to distinguish between signals feeding CGs and those feeding PGs. The first set of signals does not need any processing, while the second set of signals needs to be processed in a buffer that operated as a signal shifter, by matching its supply voltage to the desired output range. This operation can be automated by the design tool; which in turns mean that additional complexity and cost in the CAD tools and the designed circuits are expected.

If these additional problems are resolved, then the technology is very promising for logic synthesis, since it implements the XOR function in a much more efficient and compact way than CMOS. Many circuit families, including adders [38], have been reported as excellent candidates for AND-XOR-based logic minimization, instead of the classical AND-OR minimization implemented by ESPRESSO [39]. In order to maximize the benefits of the high expressive power of ambipolar CNTFET libraries, which benefits are due to the existence of XOR function embedded within the most available logic gates, it is necessary to design the minimization tool in a more efficient way, in order to make it aware of the existence of the XOR function, and to improve the efficiency of the logic mapping tool.

Another important question related to the assessments of the benefits of the designed library concerns the accuracy of the given results. This accuracy highly depends on the used device model. For the time being, there is still no SPICE-compatible compact model for ambipolar CNTFET devices. In order to simulate them in circuits, an ambipolar device is "emulated" as two n- and p-type unipolar devices in parallel. This gives an idea about the behavior of the circuit, for instance the output swing, but very accurate estimation of the delay and power consumptions cannot be derived using this model. For this reason, the simple capacitive/resistive model is utilized in order to have a first-order estimate of the delay and compare it to CMOS. This gives a good idea about the different trade-off situations explained in this chapter. The power consumption needs a more accurate modeling though, namely because it is expected to be dominated by the static power.

5.8 Chapter Contributions and Summary

This part of the work presents novel approaches to design logic circuits with ambipolar CNTFETs. The first novel family is based on dynamic logic and it can be used to generalize standard CMOS logic gates by including internal signal inversion performed by embedded XOR functions. The second novel family is based on static logic; and it addresses many issues related to dynamic logic. This static logic family is namely the first approach to design logic gates with ambipolar CNTFETs, which embeds the XOR function in a very compact way, while restoring the full-swing of the output. The proposed static logic is based on the replacement of single transistors by transmission-gates. However, many other alternate static families are proposed for the first time, realizing any possible combination between transmission-gates / pass-transistors on the one hand, and static complementary / static pseudo logic on the other hand. In this chapter, various logic design flavors with ambipolar CNTFETs are considered simultaneously for the first time. A library of different families is designed and characterized. The results are used to perform multi-level logic synthesis and to compare the different logic styles. It is therefore demonstrated that transmission-gate-based static families (both complementary and pseudo-logic) are more attractive than their pass-transistor-based counterparts. All these families have a much higher expressive power than the static CMOS family, i.e., given the same resources and topology, ambipolar CNTFETs gates are shown to implement a much larger number of logic functions than their CMOS counterparts. Moreover, the proposed efficient design combines with the benefits of the CNT technology and yield a clear improvement in terms of area, logic depth and delay compared to CMOS.

The design of regular fabrics with ambipolar CNTFETs is proposed here for the first time. Dynamic regular GNOR cells can be on-line configured as PLA planes or interconnect matrices, and they implement dynamic PLAs in a very efficient and compact way. In this approach, unlike other technologies, only one single polarity of every input is needed, leading to a reduction of the number of signals to route and making the approach very interesting for ambipolar-CNTFET-based FPGAs. On the other hand, regular static cells can be reconfigured in order to implement a large number of logic functions, that goes beyond the capabilities of CMOS cells. Besides the advantage of offering a higher expressive power than CMOS, this design approach has a predictable delay.

As it can be seen, this part represents a very innovative approach in an emerging technology that is still being discovered. The design challenges and the benefits are highly dependent on the underlying technology, which makes the approach of this chapter very different from the previous ones. This part of the book concludes this research work. The next chapter summarizes the contribution of the whole book and gives future outlooks to possible continuation of the presented work.

References

1. Kang SJ et al (2007) High-performance electronics using dense, perfectly aligned arrays of single-walled carbon nanotubes. Nat Nanotechnol 2(4):230–236
2. Zhang G, Qi P, Wang X, Lu Y, Li X, Tu R, Bangsaruntip S, Mann D, Zhang L, Dai H (2006) Selective etching of metallic carbon nanotubes by gas-phase reaction. Science 314(5801):974–977
3. Zhang J, Patil N, Mitra S (2008) Design guidelines for metallic-carbon-nanotube-tolerant digital logic circuits. In: Proceedings of DATE, pp 1009–1014
4. Patil N, Deng J, Lin A, Wong H-SP, Mitra S (2008) Design methods for misaligned and mispositioned carbon-nanotube-immune circuits. IEEE Trans Computer-Aided Des Integr Circuits Syst 27(10):1725–1736
5. Patil JZ et al (2009) Carbon nanotube circuits in the presence of carbon nanotube density variations. July 2009
6. Lin YM, Appenzeller J, Avouris P (2004) Novel carbon nanotube FET design with tunable polarity. In: IEEE International Electron Devices Meeting 2004. IEDM Technical Digest, pp 687–690
7. Lin Y-M, Appenzeller J, Knoch J, Avouris P (2005) High-performance carbon nanotube field-effect transistor with tunable polarities. IEEE Trans Nanotechnol 4(5):481–489

8. O'Connor I, Junchen L, Gaffiot F, Pregaldiny F, Lallement C, Maneux C, Goguet J, Fregonese S, Zimmer T, Anghel L, Dang T-T, Leveugle R (2007) CNTFET modeling and reconfigurable logic-circuit design. IEEE Trans Circuits Syst I: Regul Pap 54(11):2365–2379

9. O'Connor I, Liu J, Gaffiot F (2006) CNTFET-based logic circuit design. In: Proceedings of the International Conference Design and Test of Integrated Systems (DTIS), pp 46–51

10. Ben-Jamaa MH, Atienza D, Leblebici Y, Micheli GD (2008) Programmable logic circuits based on ambipolar CNFET. In: Proceedings of the Design Automation Conference (DAC), pp 339–340

11. Ben Jamaa MH, Mohanram K, De Micheli G (2009) Novel library of logic gates with ambipolar CNTFETs: opportunities for multi-level logic synthesis. In: Proceedings of DATE

12. Dresselhaus M, Dresselhaus G, Avouris P (2001) Carbon nanotubes: synthesis, structure properties and applications. Springer, Heidelberg

13. Derycke V, Martel R, Appenzeller J, Avouris P (2001) Carbon nanotube inter- and intramolecular logic gates. Nano Lett 1(9):453–456

14. Yang Q, Xiao C, Chen W, Singh AK, Asai T, Hirose A (2003) Growth mechanism and orientation control of well-aligned carbon nanotubes. Diam Relat Mater 12(9):1482–1487

15. Patil N, Lin A, Myers E, Wong HS, Mitra S (2008) Integrated wafer-scale growth and transfer of directional carbon nanotubes and misaligned-carbon-nanotube-immune logic structures. In: 2008 Symposium of the VLSI Technology, pp 205–206

16. Patil N, Deng J, Wong H-SP, Mitra S (2007) Automated design of misaligned-carbon-nanotube-immune circuits. In: DAC '07: Proceedings of the 44th annual conference on Design automation, pp 958–961

17. Deng J, Patil N, Ryu K, Badmaev A, Zhou C, Mitra S, Wong H-SP (2007) Carbon nanotube transistor circuits: circuit-level performance benchmarking and design options for living with imperfections. In: IEEE ISSCC Tech. Dig., pp 70–588

18. Bachtold A, Hadley P, Nakanishi T, Dekker C (2001) Logic circuits with carbon nanotube transistors. Science 294:1317–1320

19. Choudhury M, Yoon Y, Guo J, Mohanram K (2008) Technology exploration for graphene nanoribbon FETs. In: Proceedings of the Design Automation Conference (DAC), pp 272–277

20. Liu J, O'Connor I, Navarro D, Gaffiot F (2007) Novel CNTFET-based reconfigurable logic gate design. In: Annual ACM IEEE Design Automation Conference, pp 276–277

21. Heinze S, Tersoff J, Martel R, Derycke V, Appenzeller J, Avouris P (2002) Carbon nanotubes as Schottky barrier transistors. Phys Rev Lett 89(10):106801

22. Guo J, Datta S, Lundstrom M (2004) A numerical study of scaling issues for Schottky-barrier carbon nanotube transistors. IEEE Trans Electron Devices 51(2):172–177

23. Close GF, Yasuda S, Paul B, Fujita S, Wong H-SP (2008) A 1 GHz integrated circuit with carbon nanotube interconnects and silicon transistors. Nano Lett 8(2):706–709

24. Javey A, Guo J, Farmer DB, Wang Q, Yenilmez E, Gordon RG, Lundstrom M, Dai H (2004) Self-aligned ballistic molecular transistors and electrically parallel nanotube arrays. Nano Lett 4(7):1319–1322

25. Stanford University CNFET model (2008)

26. Goncalves NF, De Man H (1983) NORA: a racefree dynamic CMOS technique for pipelined logic structures. IEEE J Solid-State Circuits 18(3):261–266

27. Weste NHE, Harris D (2005) CMOS VLSI design: a circuits and systems perspective. Pearson/Addison Wesley, Boston

28. ABC logic synthesis tool. Available at http://www.eecs.berkeley.edu/~alanmi/abc/ for further details

29. Kheterpal V et al (2005) Design methodology for IC manufacturability based on regular logic-bricks. In: Design Automation Conference, pp 353–358

30. Mo F, Brayton RK (2002) Whirlpool PLAs: a regular logic structure and their synthesis. In: International Conference on Computer-Aided Design, pp 543–550

31. Ran Y, Marek-Sadowska M (2006) Designing via-configurable logic blocks for regular fabric. Trans Very Large Scale Integr (VLSI) Syst 14(1):1–14

32. Brockman J, Li S, Kogge P, Kashyap A, Mojarradi M (2008) Design of a mask-programmable memory/multiplier array using G4-FET technology. In: Proceedings of the Design Automation Conference (DAC), pp 337–338
33. Brown J, Taylor B, Blanton R, Pileggi L (2008) Automated testability enhancements for logic brick libraries. In: DATE 2008, pp 480–485
34. Yang S (2001) Logic synthesis and optimization benchmarks user guide version 3.0. Microelectronics Center of North Carolina, Tech. Rep., 2001
35. Sasao T (1984) Input variable assignment and output phase optimization of PLA's. IEEE Trans Comput 33(10):879–894
36. Brayton, RK Mo F (2002) Whirlpool PLAs: a regular logic structure and their synthesis. In: International Conference on Computer-Aided Design, pp 543–550
37. Schmid Y, Leblebici A (2004) Fault-tolerant PLA-style circuit design for failure-prone nanometer CMOS and quantum device technologies. In: Proceedings of the 2004 IEEE International Joint Conference on Neural Networks, 2004, vol 3, pp 1965–1969, 25–29 July 2004
38. Sasao T (1993) EXMIN2: A simplification algorithm for exclusive-OR-sum-of-products expressions for multiple-valued-input two-valued-output functions. IEEE Trans Computer-Aided Des 12(5):621–632
39. ESPRESSO. Available at http://embedded.eecs.berkeley.edu/pubs/downloads/espresso/index.htm

Chapter 6
Conclusions and Future Work

Previous chapters are dedicated to the investigation of two emerging technologies based on silicon nanowires and carbon nanotubes respectively. Both of these technologies show fundamental fabrication challenges that can be addressed by smart design approaches that leverage inherent properties of the considered technologies that do not exist in CMOS technology.

This dissertation is constructed in an interdisciplinary way. As a matter of fact, silicon nanowire technology is considered first at the fabrication level, by developing a novel fabrication technique that is suitable for the nanowire crossbar architecture. The delicate design of the decoder part in this architecture is then addressed at the system level by optimizing the encoding scheme and the decoder testing procedure.

In the last part of the work, logic design with carbon nanotube technology is investigated. The possible ambipolarity of the reported devices can be regarded as an opportunity to design logic circuits in a more efficient way than with CMOS technology. This design approach leads to regular fabrics in carbon nanotube technology that may be preferable and promising a better robustness.

In the following section, a summary of every part of the book is given. Then possible future works are proposed.

6.1 Book Summary and Contributions

Following the introduction in the first chapter, which includes a global view and a general background to regular architectures in emerging technologies, the second chapter is on the fabrication of nanowire crossbars. The third and fourth chapters are dealing with the decoder design and testing respectively. Then, the fifth chapter is on logic design with ambipolar carbon nanotubes.

In Chap. 2, an innovative approach to fabricate poly-Si nanowires, called the multi-spacer patterning technique, is introduced in order to demonstrate its ability

M. H. Ben Jamaa, *Regular Nanofabrics in Emerging Technologies*,
Lecture Notes in Electrical Engineering, 82, DOI: 10.1007/978-94-007-0650-7_6,
© Springer Science+Business Media B.V. 2011

to be extended to the fabrication of crossbars. The novelty of this process is that it uses only state-of-the-art photolithography steps, while the fabricated nanowires have their thickness and pitch far below the photolithography limit. Only CMOS processing steps are used, making the approach compatible with a CMOS process. The fabricated nanowire layers have a thickness and pitch down to 20 and 40 nm respectively, and the results are repeatable with micrometer long nanowires and a yield close to 1. A first-time demonstration of a crossbar structure fabricated exclusively with the MSPT is presented, and a crosspoint density of $\sim 10^{10}$ cm^{-2} is shown. The electrical characterization of undoped single nanowire used as poly-SiNWFETs suggests their utilization as single poly-SiNW memories. A concept of a deterministic digital decoder is proposed for the first time in this technology and it will be explored in more detail in the following chapter.

In Chap. 3, designing the decoder of crossbar circuits represents the main addressed problem. It is demonstrated that this critical part of the crossbar architecture can be made more fault-tolerant and area-efficient by deploying a multi-valued logic encoding schemes. This chapter proposes for the first time the construction rule of generalized encoding schemes in multi-valued logic. It also proposes for the first time a study of the impact of defects, locally on the code word, and globally on the code space and the crossbar yield. An analytical model that takes into account the impact of physical defects is presented in this part of the dissertation, and it can be utilized in order to investigate different trade-off situations between the encoding scheme (base, type, length) and the properties of the crossbar circuit (area, effective density, yield). It is therefore possible to explore new ways to design the decoder in order to optimize some circuit properties. This approach is applied in particular to the MSPT-decoder by defining the relation between code type, circuit variability and fabrication complexity. It is demonstrated that by optimizing the encoding scheme, it is possible to minimize either the fabrication complexity or the circuit variability or both of them, depending on the chosen encoding scheme. The best schemes are demonstrated to be the Gray-code arrangements. The arrangement of hot codes in a Gray code fashion is shown to be possible and its unique benefits in terms of reduction of variability and yield improvement are assessed for the first time.

In Chap. 4, testing the nanowire decoder part in crossbars is addressed. Generally, the nanowire decoder is not robust enough. This chapter presents for the first time a stochastic and analytical model for the impact of defects on the sensed current through the decoder. Based on this analytical model, a more accurate design of the testing unit, formed mainly by a thresholder, can be carried out in order to improve the test quality, measured as the correct test probability. Simulations performed with the developed model prove that the thresholder parameters are robust with respect to variability of design and technology parameters, while the test quality is highly dependent on these parameters. It is therefore possible to design the decoder while keeping in mind the goal of optimizing the test operation. This chapter offers a new way of design for test of crossbars, based on the optimization of the decoder parameters (such as size and code type and length) in order to improve the test quality.

In Chap. 5, new design techniques for carbon nanotube technology are investigated. Previous approaches for logic design in CNT technology considered the possible ambipolarity of the underlying device as a drawback of the technology. In this chapter, the approach is to consider this property of CNTs as an opportunity that does not exist in CMOS technology, and to assess its benefits on logic circuit design. The main novelty of this part of the dissertation is to leverage the ability of performing logic operations between the signals feeding both gates of ambipolar CNTFETs. The design with such operations is demonstrated for the first time in dynamic logic as well as in a set of static logic families including combinations of transmission-gate/pass-transistor on the one hand and complementary/pseudo logic on the other hand. This part of the work is a discovery of a natural, simple and efficient implementation of the XOR function in ambipolar CNT technology with almost no cost. The designed static families are characterized in terms of area and delay, and a benchmark of logic circuits is synthesized with the designed logic gates. This library highlights two facts. First, ambipolar CNTFET static logic has a higher expressive power than its CMOS counterpart; i.e., with the same physical resources, ambipolar CNT logic gates implement more functions than CMOS logic gates. Second, not all ambipolar CNTFET families have the same good performance. It is demonstrated through library characterization and logic synthesis, that transmission-gate based libraries are the best design approach with ambipolar CNTFETs. The proposed dynamic and static design styles can be extended to CNT regular fabrics in a straightforward way; which option is very attractive in deep sub-micron technologies.

6.2 Future Work

The discussion sections in Chaps. 2–5 highlight some open questions regarding the presented work. They also suggest many possible paths for future research work. In particular, some of the topics addressed in this work, namely spacer-based fabrication and logic design with ambipolar devices, represent novel matters that have not been explored before. There are consequently many opportunities to carry out future work especially in these fields.

On the fabrication side, the problem to be addressed is not restricted to the demonstration of the feasibility of a crossbar framework with the multi-spacer patterning technique. When it comes to the functionalization of the crossbar framework with molecular switches, many other technological problems can be investigated. For instance, the accurate control of the spacing between upper and lower nanowire layers is requited in order to allow the molecular switches to graft to poly-Si. The performance of such a complex structure formed by two nanowires with molecular switches grafted in between is a very challenging technological issue. A full crossbar circuit necessitates in addition to nanowires and molecular switches, a reliable decoder. The suggested decoder concept is a promising approach that can be investigated further in order to assess its limits in terms of maximum number of doping steps that can be tolerated within a given reliability margin.

On the system design level related to the crossbar technology, the abstract nanowire model under high variability can be enhanced by including other sources of variability to the model. It is thereby desirable to keep the same idea of abstraction that makes the model independent on the underlying nanowire fabrication technology. Such an enhanced model can be used not only to design the decoder but also to address the problem of testing it under a more accurate perspective. The sensitivity approach introduced in this work offers therefore an interesting way to classify the model parameters depending on their impact on the decoder failure and test.

Another mathematical problem that arises when we consider the system design is the generalization of the proposed optimized codes to multi-valued logic. While the generalization of Gray codes has already been considered in the past, up to now, there is no research work done with arranged hot codes besides this dissertation. The existence of such codes for small code spaces and specific logic values is demonstrated in this work by means of exhaustive search algorithms. Given the importance of these code types in terms of fault-tolerance and compactness, it is very useful to formally derive the proof (or condition) of existence of these codes in a general way, and to find their generic construction rules.

On the logic design level related to CNT technology, the proposed design approach is just a small step into a new field. When the technology and device modeling become mature enough, then many investigations at the circuit and system design levels can be further conducted. For instance the utilization of the XOR function in logic minimization is a field that has not been deeply developed, since the availability of the XOR function has never been as simple as in ambipolar CNT technology. Synthesis tools can be consequently tuned towards the specificity of the considered technology. Another interesting problem deals with the routing of signals within an FPGA where signals with one single polarity are needed. Today's routing tools for commercial FPGAs assume the existence of both signal polarities. The presented regular fabrics keep many questions open for future research. For instance, given the variety of design styles that can be implemented with the proposed regular fabrics, including FPGA and ASIC, it is interesting to consider the question of the optimal design style in the proposed approach.

These possible future works and some open question raised in the discussions of previous chapters highlight many interesting aspects in emerging technologies. It is important to assess the subtleties of the underlying emerging technology in order to discover its benefits at a higher design level and maybe to leverage some opportunities, that may not be identified as such by the technologist. It is also desirable, and luckily easier to arrange emerging technologies into regular architectures. This simplifies the task of technologists and speeds up the overall design process. Such regular architectures may depend fundamentally on the underlying technology, and may need not only a different fabrication process, but also specific design styles and methodologies that leverage their intrinsic benefits.

Index